PREFACE

In March 1987 the International Commission on Radiological Protection established a Task Group of Committee 2 "to evaluate dose per unit intake for members of the public". This task covers the development of age-dependent physical and biokinetic models and the selection of appropriate biokinetic parameters. In its first report, *ICRP Publication 56* (ICRP, 1989: Age-dependent Doses to Members of the Public from Intake of Radionuclides: Part 1), the Task Group gave dose coefficients for intakes by ingestion and inhalation of isotopes of hydrogen, carbon, strontium, zirconium, niobium, ruthenium, iodine, caesium, cerium, plutonium, americium and neptunium. For this report ingestion dose coefficients are given for isotopes of sulphur, cobalt, nickel, zinc, molybdenium, technetium, silver, tellurium and polonium using the new tissue weighting factors (w_T) given by the Commission in its 1990 recommendations. Revised ingestion dose coefficients are also included for the radioisotopes given in Part 1 using the new w_T values. In addition, ingestion dose coefficients are given for more radioisotopes. A generic model for the biokinetics of lead and the alkaline earths strontium, barium and radium has been introduced for calculating ingestion dose coefficients for radioisotopes of these elements. This model has been applied to the recalculations of the ingestion dose coefficient for ^{90}Sr, the only strontium isotope considered in Part 1. As a result of ICRP now giving new w_T values for the urinary bladder and colon, and because new information has become available on the biokinetics of plutonium, americium and neptunium in humans, the Task Group considered it to be appropriate to revise the biokinetic models for these elements given in Part 1.

The Task Group plans the following additional reports:

Ingestion dose coefficients for radioisotopes of the elements iron, selenium, antimony, thorium and uranium.
Inhalation dose coefficients for the radionuclides of *ICRP Publications 56, 67* and *69*, taking into account the new ICRP respiratory tract model.
Dose coefficients for the embryo and foetus.
Reliability of dose coefficients.

The Task Group had the following membership:

Members

A. Kaul (Chairman)	J. Inaba	D. Noßke
J. C. Barton	R. W. Leggett	J. W. Stather
K. F. Eckerman	H. Métivier	

Corresponding Members

M. R. Bailey	J. D. Harrison	H. D. Roedler†
X. Chen	L. Johansson	M. Sikov
D. Crawford-Brown	G. M. Kendall	D. M. Taylor
M. Cristy	C. R. Richmond	T. Watabe

†Deceased.

The dose coefficients in this report were computed by the Task Group on Dose Calculations. The membership of the Task Group at the time of preparation of the report was:

K. F. Eckerman (Chairman)	K. Henrichs	I. A. Likhtarev
M. Cristy	G. M. Kendall	D. Noßke
L. T. Dillman	R. W. Leggett	

During the period of preparation of this report, the membership of Committee 2 was:

1985–1989

C. B. Meinhold (Chairman)	K. F. Eckerman	N. Parmentier
W. J. Bair	A. Kaul	C. R. Richmond
X. Chen	O. Matsuoka	J. O. Snihs
R. H. Clarke	J. P. Moroni	D. M. Taylor
G. Drexler	Y. I. Moskalev†	R. H. Thomas

1989–1993

C. B. Meinhold (Chairman)	K. F. Eckerman	N. Parmentier
W. J. Bair	A. Kaul	C. R. Richmond
A. Bouville	I. A. Likhtarev	J. W. Stather
X. Chen	O. Matsuoka	D. M. Taylor
G. Drexler	H. Métivier	R. H. Thomas

The membership of the Main Commission at the time of this report was adopted was:

D. Beninson (Chairman)	W. Jacobi	G. Silini
H. Jammet (Vice chairman)	D. Li	W. K. Sinclair
H. Smith (Secretary)	J. Liniecki	B. Lindell (Emeritus)
R. H. Clarke	C. B. Meinhold	K. Z. Morgan (Emeritus)
H. J. Dunster	F. Mettler	L. S. Taylor (Emeritus)
A. K. Guskova	I. Shigematsu	

† Deceased.

GLOSSARY OF TERMS

This glossary has been revised from that given in Part 1 to include the terminology adopted by ICRP in its 1990 recommendations (ICRP, 1991).

Absorbed Dose
: the physical dose quantity defined as the quotient of the energy absorbed in an organ or tissue T by its mass (1 J kg^{-1} = 1 Gy).

Absorbed Fraction ($AF(T \leftarrow S)_i$)
: the fraction of energy emitted as a specified radiation type i in a specified source region S which is absorbed in a specified target tissue T.

Becquerel (Bq)
: the name for the SI unit of activity, 1 Bq = 1 s^{-1}.

Cells Near Bone Surfaces
: those tissues which lie within 10 μm of endosteal surfaces and bone surfaces lined with epithelium.

Committed Effective Dose
: the sum of the products of the committed organ or tissue equivalent doses and the appropriate organ or tissue weighting factors (w_T).

Committed Equivalent Dose
: the time integral of the equivalent dose rate in a particular tissue or organ that will be received by an individual following intake of radioactive material into the body. The integration time is 50 years for the adults. For children and young persons, doses are calculated to age 70 years.

Cortical Bone
: equivalent to "Compact Bone" in *ICRP Publication 20* (ICRP, 1973), i.e. any bone with a surface/volume ratio less than 60 cm^2 cm^{-3}; in Reference Man it has a mass of 4000 g.

Dose Coefficient
: committed tissue equivalent dose per unit intake or committed effective dose per unit intake (Sv Bq^{-1}).

Effective Dose
: the effective dose is the sum of the weighted equivalent doses in all the tissues and organs of the body.

Equivalent Dose
: the tissue or organ absorbed dose (Gy or J kg^{-1}) multiplied by the radiation weighting factor (w_R).

Fractional Absorption in the Gastrointestinal Tract
the fraction of an element entering the gastrointestinal tract which reaches body fluids is termed the f_1 value.

Prompt Excretion
removal of activity directly from blood to urinary bladder or to the gastrointestinal tract with the clearance half-time assigned to blood.

Radiation Weighting Factor (w_R)
the radiation weighting factor is a dimensionless factor to derive the equivalent dose from the absorbed dose averaged over a tissue or organ and is based on the quality of radiation.

Red Bone Marrow (active)
the component of marrow which contains the bulk of the haematopoietic stem cells.

Reference Man
a person with the anatomical and physiological characteristics defined in the report of the ICRP Task Group on Reference Man (ICRP, 1975).

Sievert (Sv)
the name for the SI unit of equivalent dose and effective dose:
$1\ \text{Sv} = 1\ \text{J kg}^{-1}$.

Source Region (S)
region within the body containing the radionuclide. The region may be an organ, a tissue, or the contents of the gastrointestinal tract or urinary bladder.

Specific Effective Energy ($SEE(T \leftarrow S)_i$)
the energy (MeV), suitably modified for radiation weighting factor, imparted per gram of a target tissue T as a consequence of the emission of a specified radiation (i) from a transformation occurring in source region S.

Target Tissue
tissue or organ in which radiation is absorbed.

Tissue Weighting Factor (w_T)
the tissue weighting factors are dimensionless factors to derive the effective dose from the equivalent dose. They are based on the different sensitivities of tissues to radiation.

Trabecular Bone
equivalent to "Cancellous Bone" in *ICRP Publication 20* (ICRP, 1973), i.e. any bone with a surface/volume ratio greater than 60 $cm^2\ cm^{-3}$; in Reference Man it has a mass of 1000 g.

References

ICRP (1973) *Alkaline Earth Metabolism in Adult Man*, ICRP Publication 20. Pergamon Press, Oxford.
ICRP (1975) *Report of the Task Group on Reference Man*, ICRP Publication 23, pp. 390–391. Pergamon Press, Oxford.
ICRP (1991) *1990 Recommendations of the International Commission on Radiological Protection Publication 60.* Annals of the ICRP **21**(1–3), 1991.

INTRODUCTION

(1) The present report is Part 2 of a report published as *ICRP Publication 56*, on age-dependent dose coefficients to members of the public from intakes by ingestion of radioisotopes of the following elements: sulphur, cobalt, nickel, zinc, strontium, molybdenum, technetium, silver, tellurium, barium, lead, polonium and radium. Also included are ingestion dose coefficients for radioisotopes of the following elements of Part 1: hydrogen, carbon, zirconium, niobium, ruthenium, iodine, caesium, cerium, plutonium, americium and neptunium. This report gives parameters for the tissue distribution and retention of these elements together with data on urinary and faecal excretion. Information on excretion pathways is required because of the need to estimate doses to the urinary bladder and colon (ICRP, 1991a).

(2) The biokinetic model for strontium used in Part 1 (ICRP, 1989) was considered suitable for application to ^{90}Sr, the only isotope of strontium considered in that report, but an alkaline earth model developed since the publication of Part 1 is expected to be more appropriate for application to the shorter-lived isotope, ^{89}Sr, considered in this report. This model described in Appendix A is considered to give a better representation of bone physiology. For consistency, revised dose coefficients for ^{90}Sr based on the new model are also given.

(3) The biokinetic models used for calculating dose coefficients for plutonium, americium and neptunium have been modified to include information on their behaviour in humans which was not available when the models for these elements in Part 1 were being developed. It has also been necessary to depict excretion pathways explicitly in these models. The revised biokinetic models for plutonium, americium, and neptunium are described in Appendix B.

(4) Dose coefficients for the radioisotopes of Part 1 and Part 2 for intakes by inhalation will be recalculated using the new Human Respiratory Tract Model (ICRP, 1994).

(5) Dose coefficients have been calculated for radioisotopes of the elements covered by this report which are expected to be released into the environment as a result of human activities and considered to be of significance for environmental radiation protection purposes.

(6) The age-dependent dose coefficients for selected radioisotopes of these elements are based on reviews of biokinetic data and the general dosimetric models described in Part 1 (ICRP, 1989). For radioisotopes of the alkaline earths and lead, new age-dependent biokinetic models for uptake and retention in the skeleton and other tissues were adopted by the Task Group. The models are described in detail in Appendix A. The movement of elements between body compartments in a recycling model is best described by a series of transfer rates. The parameters are derived from model fits to experimental data and are given to a sufficient precision to maintain numerical accuracy.

(7) For elements retained in the skeleton for an extended period but for which no retention functions are available a biological half-time corresponding to the age-dependent turnover rate of cortical bone has been assumed. Half-times adopted for radionuclides retained in bone volume are 100, 300, 500, 1000, 1500 and 10,000 days for the 3-month-old infant, 1-, 5-, 10-, 15-year-old children, and adults, respectively. This approach has also been applied to elements in Part 1 and the modified biokinetic data are given in Appendix C.

(8) Where no clear evidence on age dependence of organ distribution and retention appeared to be available, the biokinetic data for adults were adopted for infants and children. This assumption was made in Part 1 (ICRP, 1989) and is usually expected to lead to an overestimate of the dose coefficients.

(9) If no relevant biokinetic data were found for humans, appropriate data were based on animal experiments.

(10) Many elements incorporated into food may be more readily absorbed from the gastrointestinal tract than inorganic forms of these elements (ICRP, 1989). For elements with f_1 values in *ICRP Publication 30* (ICRP, 1979, 1980, 1981) greater than 0.5, and for which there is no specific information on absorption from food, the approach adopted here is to apply an f_1 value of 1.

(11) Absorption of radionuclides tends to be greatest in the newborn but the results of animal studies suggest that the gut transfer progressively decreases with age, reaching adult values by about the time of weaning in most cases. As described in Part 1 where no human or animal data are available, the following approach, based on the general body of animal data supported by limited human data, was adopted. For elements with a fractional absorption in the adult of 0.001 or less, a value 10 times the value for the adult is assumed. For fractional absorption values between 0.01 and 0.5 in the adult, an increase by a factor of 2 is assumed for the 3 month old infant. This approach is based on the recommendations of an NEA Expert Group (NEA/OECD, 1988). For the group of elements with an assumed increase of the f_1 value of a factor 10 the actinides are examples for which the experimental evidence is well established (ICRP, 1986, 1989). For the group of elements with an assumed increase in the f_1 value of a factor 2 niobium is an example which is supported by data from rats and guinea pigs (ICRP, 1989). For absorption values greater than 0.5 in the adult, the approach adopted here is to assume complete absorption by 3 month old infants ($f_1 = 1$).

(12) Some radionuclides are assumed to be completely absorbed from the gastrointestinal tract. For such radionuclides, all the activity leaving the stomach is taken to enter the transfer compartment. For other nuclides, a fraction of the ingested activity enters the transfer compartment from the small intestine, the remaining fraction traverses the gastrointestinal tract without being absorbed. If complete absorption is indicated at some, but not all ages, or if complete absorption is indicated for some, but not all members of a decay chain, then for technical reasons calculations are performed assuming that absorption occurs from the small intestine with 0.01 of the ingested material entering the lower segments of the tract; the fraction absorbed being 0.99. For loss of activity from the transfer compartment, a half-time of 0.25 days has been adopted (ICRP, 1979), unless specified otherwise in the biokinetic data tables.

(13) Activity within the systemic circulation is taken to be distributed over a mass which excludes the masses of the contents of the gastrointestinal tract, the urinary bladder and the gall bladder. Consequently, the specific activity of a radionuclide stated to be uniformly distributed in the total body of the adult in this report is based on a mass of 68.8 kg rather than the 70 kg value used earlier. For example, this change in definition results in an increase, albeit small, in the committed equivalent dose for tritium and carbon given here over the values in Part 1 (ICRP, 1989).

(14) The tissue weighting factors w_T used for calculating the effective dose coefficients are those given by ICRP in its 1990 recommendations (ICRP, 1991a); see Table I.-1. Dose coefficients are given in this report for both the ovaries and testes, the higher of which is applied to the gonad weighting factor for the calculation of effective

Table I.-1. Tissue weighting factors[1]

Tissue or organ	Tissue weighting factor w_T
Gonads	0.20
Bone marrow (red)	0.12
Colon	0.12
Lung	0.12
Stomach	0.12
Bladder	0.05
Breast	0.05
Liver	0.05
Oesophagus	0.05
Thyroid	0.05
Skin	0.01
Bone surface	0.01
Remainder[2]	0.05

[1] The values have been developed from a reference population of equal numbers of both sexes and a wide range of ages. In the definition of effective dose they apply to workers, to the whole population, and to either sex.

[2] For purposes of calculation, the remainder is composed of the following nine additional tissues and organs: adrenals, brain, small intestine, kidney, muscle, pancreas, spleen, thymus and uterus. The list includes organs which are likely to be selectively irradiated. Some organs in the list are known to be susceptible to cancer induction. If other tissues and organs subsequently become identified as having a significant risk of induced cancer they will then be included either with a specific w_T or in this additional list constituting the remainder. The latter may also include other tissues or organs selectively irradiated.

dose. The Commission has clarified application of the colon weighting factor given in Table I.-1. The weighting factor is to be applied to the mass average of the equivalent dose in the walls of the upper and lower large intestine of the gastrointestinal tract. The upper large intestine is no longer included in the remainder tissues. Age-specific masses of the walls of the gastrointestinal tract are given in Table I.-1 of Part 1 (ICRP, 1989). Since the relative masses of the walls of the upper large intestine (ULI) and lower large intestine (LLI) are independent of age the equivalent dose to the colon H_{colon} is given as

$$H_{colon} = 0.57\, H_{ULI} + 0.43\, H_{LLI}$$

where H_{ULI} and H_{LLI} are the equivalent doses in the walls of the ULI and LLI, respectively.

In the exceptional case in which the most exposed remainder tissue or organ receives the highest committed equivalent dose of all organs, a weighting factor of 0.025 (half of remainder) is applied to that tissue or organ and 0.025 (half of remainder) to the mass average committed equivalent dose in the rest of the remainder tissues and organs.

(15) In the 1990 recommendations (ICRP, 1991a) the urinary bladder is given an explicit w_T value as shown in Table I.-1. Consequently, a urinary bladder model is formulated for calculating the additional equivalent dose to the bladder wall from activity in the urine.

The radiation dose to the bladder wall involves a complex relationship between urine flow rate, voiding, and urine volume present in the bladder when the radionuclide entered the bladder and is critically dependent on the geometrical relationship between the wall of the bladder and its contents. For radiation protection purposes, such a model was developed by Snyder and Ford (1976) and was extended by Smith *et al.* (1982) to administered radiopharmaceuticals. This model was adopted in *ICRP Publication 53* (ICRP, 1987) and is modified in this report to facilitate its application to long-lived radionuclides. The bladder is taken to be of fixed size containing 15, 25, 65, 75, 85 and 115 ml of urine in the 3-month-old, 1-, 5-, 10-, 15-year-old children, and adults, respectively. These volumes represent the average content of the bladder during the time period between voids. The rate at which radionuclides enter the bladder is based on their elimination rates from body tissues and the urine to faecal excretion ratio given with the biokinetic data. For some elements, the biokinetic data directly address excretion; e.g. the iodine and lead of the alkaline-earth and actinide elements. While the urine to faecal excretion ratio for an element may vary with time, a constant value is judged to be adequate for the present dosimetric purposes. It is not intended that the stated ratios be used for the interpretation of bioassay measurements where temporal variations in the ratio may be important. The number of voids per day for the 3-month and 1-year-old are taken as 20 and 16, respectively (Goellner *et al.*, 1981). For all other ages, 6 voids per day are assumed (Larsson and Victor, 1988; Bloom *et al.*, 1993). The volume of the bladder contents, noted above, correspond to one-half the volume of urine excreted in a void (ICRP, 1975). To represent the kinetics of the bladder in terms of first order processes, the rate of elimination from the bladder is taken to be twice the number of voids per day. That is, the elimination rates from the bladder would be equivalent to 40, 32, 12, 12, 12 and 12 day^{-1} for the 3-month-old infants, 1-, 5-, 10-, 15-year-old children, and the adults, respectively.

(16) The Commission has assigned an explicit w_T to the oesophagus in Table I.-1. The biokinetic model presented here adopted from *ICRP Publication 30* (ICRP, 1979) contains no information on uptake and retention of radionuclides in the oesophagus. Since the transit time of materials through the oesophagus is quite rapid, only the dose from penetrating radiation emitted from other source organs needs to be considered. In the absence of a dosimetric model for the oesophagus the dose of the thymus is used as a surrogate (ICRP, 1991b).

(17) The activity in the upper and lower large intestine at times following the intake of a radionuclide includes activity being eliminated from the systemic circulation. Except where otherwise indicated, it is assumed that systemic activity being lost in faeces was secreted into the upper large intestine.

(18) Dose coefficients for intakes by ingestion by 3-month-old infants, 1-, 5-, 10- and 15-year-old children and adults are given in this report. In the calculations of the activity in source regions of the body following intakes at these ages, continuous changes with age in the transfer rates governing the distribution and retention of the activity are obtained by linear interpolation according to age. This applies to the transfer of activity from the small intestine to body fluid which is generally stated in terms of the f_1 value. For applications to other ages the Task Group considered that tissue doses can be estimated by applying the age-specific dose coefficients to the age ranges given below:

3 months: infants from 0 to 12 months of age

1 year: from 1 year to 2 years
5 years: more than 2 years to 7 years
10 years: more than 7 years to 12 years
15 years: more than 12 years to 17 years
adult: more than 17 years.

(19) As in Part 1, a single reference subject is used to represent each age-group. Generally, parameter values for males have been adopted because of the availability of biokinetic data. Where there are known gender differences in the biokinetics of an element this is noted in the relevant section of the biokinetic data.

(20) Except where otherwise indicated, adulthood is assumed to be reached at age 20 years, and dose coefficients for the adult are based on equivalent dose rates received over a 50-year period following acute intake at age 20 years. Exceptions are made for the alkaline-earth elements, lead, plutonium, americium and neptunium. For these elements, the transfer rates for the adult apply to ages ≥ 25 years, and dose coefficients for the adult are based on equivalent dose rates received over a 50-year period following acute intake at age 25 years. This approach is used because some of the transfer rates in the biokinetic models for these elements are equated with bone formation rates, which are expected to remain elevated during the early part of the third decade of life.

(21) The doses calculated in this report and in Part 1 are for acute intakes. For chronic intakes doses may be less than those calculated here where growth is substantial during the period of intake.

References

Bloom, D. A., Seeley, W. W., Ritchey, M. L. and McGuire, E. J. (1993) Toilet habits and continence in children: An opportunity sampling in search of normal parameters. *J. Urol.* **149**, 1087–1090.

Goellner, M. H., Ziegler, E. E. and Fomon, S. J. (1981) Urination during the first three years of life. *Nephron* **28**, 174–178.

ICRP (1975) *Report of the Task Group on Reference Man*, ICRP Publication 23. Pergamon Press, Oxford.

ICRP (1979) *Limits for Intakes of Radionuclides by Workers*, ICRP Publication 30, Part 1. Pergamon Press, Oxford.

ICRP (1980) *Limits for Intakes of Radionuclides by Workers*, ICRP Publication 30, Part 2. Pergamon Press, Oxford.

ICRP (1981) *Limits for Intakes of Radionuclides by Workers*, ICRP Publication 30, Part 3. Pergamon Press, Oxford.

ICRP (1986) *The Metabolism of Plutonium and Related Elements*, ICRP Publication 48. Pergamon Press, Oxford.

ICRP (1987) *Radiation Dose to Patients from Radiopharmaceuticals*, ICRP Publication 53. Pergamon Press, Oxford.

ICRP (1989) *Age-dependent Doses to Members of the Public from Intake of Radionuclides: Part 1*, ICRP Publication 56. Pergamon Press, Oxford.

ICRP (1991a) *1990 Recommendations of the International Commission on Radiological Protection*, ICRP Publication 60. Pergamon Press, Oxford.

ICRP (1991b) *Annual Limits on Intake of Radionuclides by Workers Based on the 1990 Recommendations*, ICRP Publication 61. Pergamon Press, Oxford.

ICRP (1994) *Human Respiratory Tract Model for Radiological Protection*, ICRP Publication 66.

Larsson, G. and Victor, A. (1988) Micturition patterns in a healthy female population, studied with a frequency/volume chart. *Scand. J. Urol. Nephrol. Supp.* **114**, 53–57.

NEA/OECD (1988) Committee on radiation protection and public health. *Report of an Expert Group on Gut Transfer Factors*. NEA/OECD Report, Paris.

Smith, T., Veall, N. and Wootten, R. (1982) Bladder wall dose from administered radiopharmaceuticals: The effect on variation in urine flow rate, voiding interval and initial bladder content. *Radiat. Prot. Dosim.* **2**, 183–189.

Snyder, W. S. and Ford, M. R. (1976) Estimation of doses to the urinary bladder and to the gonads. In: *Radiopharmaceutical Dosimetry Symp.* (Proc. Conf. Oak Ridge, Tennessee, April 1976). HEW Publication (FDA 76-8044), pp. 313–349. Department of Health, Education and Welfare, Bureau of Radiological Health, Rockville, MD.

COMPUTATION OF AGE-DEPENDENT EFFECTIVE DOSE COEFFICIENTS

(22) Part 1 of (ICRP, 1989) outlined the computational formulations of the age-dependent dosimetric quantities. Some additional discussion is necessary in implementing, in an age-dependent manner, the effective dose quantity defined in the 1990 recommendations (ICRP, 1991a).

(23) In Part 1 the equivalent dose rate at age t in target organ or tissue T due to an acute intake of a radionuclide by an individual of age t_0 at the time of intake, $\dot{H}_T(t, t_0)$, was expressed as:

$$\dot{H}_T(t, t_0) = C \sum_S q_S(t) \mathrm{SEE}(\mathrm{T} \leftarrow \mathrm{S}; t)$$

where $q_S(t)$ is the activity of the radionuclide in source region S at age t and $\mathrm{SEE}(\mathrm{T} \leftarrow \mathrm{S}; t)$ is the specific effective energy deposited in target organ T per nuclear transformation of the radionuclide in source region S at age t, and C is any numerical constant required by the units of q and SEE. The total committed equivalent dose in T accumulated by age 70 years due to the intake at age t_0, $H_T(t_0)$, is

$$H_T(t_0) = \int_{t_0}^{70} \dot{H}_T(t, t_0)\, \mathrm{d}t = C \int_{t_0}^{70} \sum_S q_S(t) \mathrm{SEE}(\mathrm{T} \leftarrow \mathrm{S}; t)\, \mathrm{d}t.$$

When a radionuclide decays to a decay product which itself is radioactive, an exactly similar set of equations applies to doses from the decay products. The committed effective dose accumulated by age 70 years due to the intake at age t_0, $E(t_0)$, is

$$E(t_0) = \sum_{T=1}^{T=12} w_T H_T(t_0) + w_{rem} H_{rem}(t_0)$$

where the summation extends over 12 organs and tissues in Table I.-1 assigned explicit weighting factors, w_{rem} is the weighting factor for the remainder tissues, and $H_{rem}(t_0)$ is the equivalent dose accumulated by age 70 years in remainder tissues. The equivalent dose in the remainder, $H_{rem}(t_0)$, is formulated in terms of the average equivalent dose in the tissues comprising the remainder (ICRP, 1991a). For the adult, $H_{rem}(t_0)$ can be evaluated directly from the dose equivalent in these tissues since their masses are independent of age; see paragraphs 6 and 7 of *ICRP Publication 61* (ICRP, 1991b). For children, however, the evaluation of $H_{rem}(t_0)$ must reflect the age-dependence of the organ masses.

(24) The computational form of the equivalent dose rate in the remainder depends upon whether any organ or tissue of the remainder experiences an equivalent dose in excess of the committed equivalent dose in organs and tissues given explicit weighting factors (ICRP, 1991a,b). Let H_{max} denote the maximum committed equivalent dose among all organs with explicit weighting factors and $H_{T'}$, the maximum committed equivalent dose among organs of the remainder. The equivalent dose rate in the remainder tissues, $\dot{H}_{rem}(t, t_0)$, is then

$$\dot{H}_{\text{rem}}(t, t_0) = \begin{cases} \dfrac{\sum_{T=1}^{T=9} m_T(t)\dot{H}(t, t_0)}{\sum_{T=1}^{T=9} m_T(t)}, & \text{if } H_{T'} \leq H_{\max} \\[2ex] 0.5\left[\dfrac{\sum_{T=1(T\neq T')}^{T=9} m_T(t)\dot{H}_T(t, t_0)}{\sum_{T=1(T\neq T')}^{T=9} m_T(t)} + \dot{H}_{T'}(t, t_0)\right], & \text{if } H_{T'} > H_{\max} \end{cases}$$

where the summation extends over the nine individual organs and tissues or the remainder. In implementing the above equation, the organ masses, $m_T(t)$, are obtained by interpolation of the data in Table 1.-1 of Part 1 (ICRP, 1989). The equivalent dose accumulated by age 70 years in the remainder tissue due to an intake at age t_0 is

$$H_{\text{rem}} = \int_{t_0}^{70} \dot{H}_{\text{rem}}(t, t_0)\, \mathrm{d}t.$$

In the case of the adult, the organ masses are assumed to be independent of age and thus the formulation is identical to that in *ICRP Publication 61* (ICRP, 1991a).

References

ICRP (1989) *Age-dependent Doses to Members of the Public from Intake of Radionuclides: Part 1*, ICRP Publication 56. Pergamon Press, Oxford.

ICRP (1991a) *1990 Recommendations of the International Commission on Radiological Protection*, ICRP Publication 60. Pergamon Press, Oxford.

ICRP (1991b) *Annual Limits on Intake of Radionuclides by Workers based on the 1990 Recommendations*, ICRP Publication 61. Pergamon Press, Oxford.

PREFACE

(25) Based upon a review of biokinetic data for selected elements, age-dependent reference values of uptake, distribution and excretion have been adopted for calculating dose coefficients for different ages at intake. In the present report the Task Group has calculated ingestion dose coefficients for the following radionuclides:

^{35}S; ^{57}Co; ^{58}Co; ^{60}Co; ^{59}Ni; ^{63}Ni; ^{65}Zn; ^{89}Sr; ^{90}Sr; ^{99}Mo; ^{99}Tc; ^{99m}Tc; ^{108m}Ag; ^{110m}Ag; ^{127m}Te; ^{129m}Te; ^{131m}Te; ^{132}Te; ^{133}Ba; ^{140}Ba; ^{210}Pb; ^{210}Po; ^{224}Ra; ^{226}Ra; ^{228}Ra.

(26) In addition in Appendix C revised biokinetic data and dose coefficients are given for the following radioisotopes of elements of Part 1:

^{3}H; ^{14}C; ^{95}Zr; ^{95}Nb; ^{103}Ru; ^{106}Ru; ^{125}I; ^{129}I; ^{131}I; ^{132}I; ^{134}Cs; ^{136}Cs; ^{137}Cs; ^{141}Ce; ^{144}Ce; ^{238}Pu; ^{239}Pu; ^{240}Pu; ^{241}Pu; ^{241}Am; ^{237}Np; ^{239}Np.

1. SULPHUR

Uptake to blood

(27) *Adults.* Volwiler *et al.* (1955) reported that the fractional absorption of sulphur (S) in adult men was greater than 0.6 for organic compounds. Minski and Vennart (1971) measured absorption in rats and obtained a mean value for the f_1 for ^{35}S 1-methionine of 0.92 ± 0.18 (± 1 SD) and Dziewiatkowski (1945, 1949) found the f_1 value for ^{35}S in sodium sulphide and sodium sulphate administered to rats was approximately equal to 1. Later Dziewiatkowski (1962) reported that elemental sulphur is less well absorbed than inorganic compounds of the element. In *ICRP Publication 30* (ICRP, 1980) the f_1 value given for inorganic compounds of sulphur was taken to be 0.8 and a value of 0.1 was adopted for sulphur in its elemental form. Previously ICRP (1959) had suggested an f_1 value of 1.

(28) Because there appears to be almost complete absorption of ^{35}S ingested as amino acids or other organic compounds, almost complete absorption of ^{35}S incorporated in foodstuffs is expected to occur. For sulphur present in the diet, an f_1 value of 1 is adopted here following the general approach described in paragraph (10) of the Introduction. When a large proportion of dietary sulphur is present in inorganic forms a lower f_1 value could be adopted based on specific information on absorption.

(29) *Children.* There are few data available on the absorption of sulphur in infants and children. Shohl and Sato (1923) showed that an 8-month-old baby lost about 3% of its dietary sulphur intake in faeces, which indicates absorption of at least 97%. This value is consistent with available data for adults. However, Wright *et al.* (1960) reported sulphur balance studies which indicated that 68–90% of dietary sulphur was absorbed by pre-adolescent girls, apparently depending on dietary nitrogen levels.

(30) According to the general approach described in paragraph (11) of the Introduction, an f_1 value of 1 is adopted here for infants and children, the same as the value adopted for adults.

Distribution and retention

(31) *Adults.* Minski and Cleary (1966) reported measurements of the concentration of stable sulphur in a number of human tissues and organs. The concentrations in most tissues, including testes, were in the range 1–2 mg g^{-1}, although that for cartilage was higher, 5.5 mg g^{-1}, and that in bone marrow was less, 0.7 mg g^{-1}. Minski and Vennart (1971) studied the distribution and retention of ^{35}S in rats up to 130 days after injection of ^{35}S 1-methionine and up to 8 days after injection of ^{35}S sodium sulphate. For the organic compound the distribution of ^{35}S among most tissues was fairly uniform; with the exception of the femur which had the lowest concentration, only about a factor of 3 separated the organs with the highest concentration (muscle) and lowest concentration (testes). For the inorganic compound the range of concentrations between different tissues was about a factor of 20 between the highest (cartilage) and lowest (muscle). Maass *et al.* (1949) reported that in rats there was no selective localization of ^{35}S in any tissues up to 47 h after administration of labelled methionine, although concentrations in individual tissues varied by a factor of about 10. The muscle, testes and brain of the rat had significantly lower concentrations than other tissues. The results obtained after 4 and

47 h suggest a tendency for tissues to approach a similar level of activity. Taken together these results suggest that for the purposes of dosimetry, it is appropriate to assume that ^{35}S is distributed uniformly throughout all body tissues.

(32) The retention of ^{35}S in the body appears to depend on the chemical form administered, with inorganic forms lost more rapidly than organic forms. Woodard *et al.* (1976) reported that 70–90% of ^{35}S intravenously injected as sodium sulphate in man was excreted in the urine in the first 3 days. Similar results in rats were reported by Minski and Vennart (1971) who also reported that for rats given ^{35}S 1-methionine the rate of loss was about 10 times slower. Walser *et al.* (1953) reported that in humans given intravenous injections of ^{35}S in the form of dilute H_2SO_4, about 5% of the administered activity was excreted within about 0.3 h and about 50% within 4–9 h. These human data suggest that for inorganic forms of sulphur there is a rapid phase of clearance with a biological half-time of about 0.3 days followed by a slower phase with a half-time of at least 7 days, and possibly as much as 80 days (ICRP, 1980). However, for intakes of dietary sulphur (0.8 g day^{-1}) (ICRP, 1975), two such components are insufficient to explain the total body content of sulphur given for Reference Man (140 g) and a third long-term component of retention has been postulated. In *ICRP Publication 30* (ICRP, 1980) it was proposed that for sulphur leaving the transfer compartment with a half-time of 0.25 days a fraction of 0.8 should be assumed to be directly excreted and fractions 0.15 and 0.05 should be assumed to be distributed uniformly throughout all organs and tissues of the body, and to be retained with biological half-times of 20 and 2000 days, respectively. These retention parameters are reasonably consistent with the available data and are adopted here for the calculation of dose coefficients for adults in the population ingesting inorganic forms of the element as may occur in drinking water.

(33) For sulphur incorporated into amino acids and other organic forms of sulphur, rapid urinary excretion of a major fraction of sulphur from blood would not be expected. Results obtained by Minski and Vennart (1971) were consistent with rapid urinary excretion of sulphate from the blood compared with retention of methionine. Amino acids are efficiently reabsorbed in the proximal tubules of the kidney. Taking account of these data and dietary intake and body content data (ICRP, 1975), Vennart and Ash (1976) proposed that organic sulphur in food should be assumed to be completely absorbed and uniformly distributed in body tissues with a single component of retention. In this report an f_1 of 1.0 and whole body retention with a half-time of 140 days are assumed for sulphur in food. In this report, a whole body retention with a half-time of 140 days is assumed for sulphur in food. A urinary to faecal excretion ratio of 9:1 is assumed for all forms of sulphur that has entered the transfer compartment.

(34) *Children.* The available data for infants and children are consistent with more rapid turnover of sulphur in the body (Vennart and Ash, 1976). However, the distribution of sulphur may be different in young children with greater uptake into chondroitin sulphate in collagen (Dziewiatkowski, 1962). Adult retention parameters are therefore conservatively applied in this report to infants and children.

Dose coefficients

(35) The biokinetic data are summarized in Tables 1.-1 and 1.-2. The derived dose coefficients are given in Tables 1.-3 and 1.-4.

Table 1.-1. Biokinetic data for inorganic sulphur

Age	f_1	Distribution (%)			Biological half-time (days)	
		Prompt excretion	Total body		Total body	
			Comp. A	Comp. B	Comp. A	Comp. B
3 months	1	80	15	5	20	2000
1 year	1	80	15	5	20	2000
5 years	1	80	15	5	20	2000
10 years	1	80	15	5	20	2000
15 years	1	80	15	5	20	2000
Adult	1	80	15	5	20	2000

A urinary to faecal excretion ratio of 9:1 is assumed for sulphur that has entered the transfer compartment.

Table 1.-2. Biokinetic data for organic sulphur

Age	f_1	Distribution (%)	Biological half-time (days)
		Total body	Total body
3 months	1	100	140
1 year	1	100	140
5 years	1	100	140
10 years	1	100	140
15 years	1	100	140
Adult	1	100	140

A urinary to faecal excretion ratio of 9:1 is assumed for sulphur that has entered the transfer compartment.

Table 1.-3

Ingestion Dose Coefficients (Equivalent Dose (Sv/Bq) to Age 70 Years) for Inorganic Sulphur (S-35 T1/2 = 87.44 d)

Age at intake	3 Months	1 Year	5 Years	10 Years	15 Years	Adult
Adrenals	9.5E-10	6.6E-10	3.3E-10	2.0E-10	1.2E-10	9.6E-11
Bladder Wall	1.4E-09	1.0E-09	6.8E-10	5.0E-10	3.8E-10	2.9E-10
Bone Surfaces	9.5E-10	6.6E-10	3.3E-10	2.0E-10	1.2E-10	9.6E-11
Brain	9.5E-10	6.6E-10	3.3E-10	2.0E-10	1.2E-10	9.6E-11
Breast	9.5E-10	6.6E-10	3.3E-10	2.0E-10	1.2E-10	9.6E-11
GI-Tract						
St Wall	1.7E-09	1.1E-09	5.2E-10	3.1E-10	1.9E-10	1.5E-10
SI Wall	9.7E-10	6.7E-10	3.4E-10	2.0E-10	1.2E-10	9.8E-11
ULI Wall	2.0E-09	1.3E-09	6.6E-10	3.9E-10	2.3E-10	1.8E-10
LLI Wall	3.8E-09	2.5E-09	1.3E-09	7.5E-10	4.3E-10	3.5E-10
Kidneys	9.5E-10	6.6E-10	3.3E-10	2.0E-10	1.2E-10	9.6E-11
Liver	9.5E-10	6.6E-10	3.3E-10	2.0E-10	1.2E-10	9.6E-11
Lungs	9.5E-10	6.6E-10	3.3E-10	2.0E-10	1.2E-10	9.6E-11
Muscle	9.5E-10	6.6E-10	3.3E-10	2.0E-10	1.2E-10	9.6E-11
Ovaries	9.5E-10	6.6E-10	3.3E-10	2.0E-10	1.2E-10	9.6E-11
Pancreas	9.5E-10	6.6E-10	3.3E-10	2.0E-10	1.2E-10	9.6E-11
Red Marrow	9.5E-10	6.6E-10	3.3E-10	2.0E-10	1.2E-10	9.6E-11
Skin	9.5E-10	6.6E-10	3.3E-10	2.0E-10	1.2E-10	9.6E-11
Spleen	9.5E-10	6.6E-10	3.3E-10	2.0E-10	1.2E-10	9.6E-11
Testes	9.5E-10	6.6E-10	3.3E-10	2.0E-10	1.2E-10	9.6E-11
Thymus	9.5E-10	6.6E-10	3.3E-10	2.0E-10	1.2E-10	9.6E-11
Thyroid	9.5E-10	6.6E-10	3.3E-10	2.0E-10	1.2E-10	9.6E-11
Uterus	9.5E-10	6.6E-10	3.3E-10	2.0E-10	1.2E-10	9.6E-11
Remainder	9.5E-10	6.6E-10	3.3E-10	2.0E-10	1.2E-10	9.6E-11
Effective Dose	1.3E-09	8.7E-10	4.4E-10	2.7E-10	1.6E-10	1.3E-10

GI-Tract Gastrointestinal Tract
St Stomach
SI Small Intestine
ULI Upper Large Intestine
LLI Lower Large Intestine

Table 1.-4

Ingestion Dose Coefficients (Equivalent Dose (Sv/Bq) to Age 70 Years) for Organic Sulphur S-35 (T1/2 = 87.44 d)

Age at intake	3 Months	1 Year	5 Years	10 Years	15 Years	Adult
Adrenals	7.4E-09	5.2E-09	2.6E-09	1.6E-09	9.3E-10	7.6E-10
Bladder Wall	7.6E-09	5.4E-09	2.8E-09	1.7E-09	1.0E-09	8.4E-10
Bone Surfaces	7.4E-09	5.2E-09	2.6E-09	1.6E-09	9.3E-10	7.6E-10
Brain	7.4E-09	5.2E-09	2.6E-09	1.6E-09	9.3E-10	7.6E-10
Breast	7.4E-09	5.2E-09	2.6E-09	1.6E-09	9.3E-10	7.6E-10
GI-Tract						
St Wall	8.2E-09	5.6E-09	2.8E-09	1.7E-09	1.0E-09	8.1E-10
SI Wall	7.5E-09	5.2E-09	2.6E-09	1.6E-09	9.3E-10	7.6E-10
ULI Wall	7.9E-09	5.5E-09	2.8E-09	1.7E-09	9.8E-10	8.0E-10
LLI Wall	8.7E-09	6.1E-09	3.1E-09	1.8E-09	1.1E-09	8.8E-10
Kidneys	7.4E-09	5.2E-09	2.6E-09	1.6E-09	9.3E-10	7.6E-10
Liver	7.4E-09	5.2E-09	2.6E-09	1.6E-09	9.3E-10	7.6E-10
Lungs	7.4E-09	5.2E-09	2.6E-09	1.6E-09	9.3E-10	7.6E-10
Muscle	7.4E-09	5.2E-09	2.6E-09	1.6E-09	9.3E-10	7.6E-10
Ovaries	7.4E-09	5.2E-09	2.6E-09	1.6E-09	9.3E-10	7.6E-10
Pancreas	7.4E-09	5.2E-09	2.6E-09	1.6E-09	9.3E-10	7.6E-10
Red Marrow	7.4E-09	5.2E-09	2.6E-09	1.6E-09	9.3E-10	7.6E-10
Skin	7.4E-09	5.2E-09	2.6E-09	1.6E-09	9.3E-10	7.6E-10
Spleen	7.4E-09	5.2E-09	2.6E-09	1.6E-09	9.3E-10	7.6E-10
Testes	7.4E-09	5.2E-09	2.6E-09	1.6E-09	9.3E-10	7.6E-10
Thymus	7.4E-09	5.2E-09	2.6E-09	1.6E-09	9.3E-10	7.6E-10
Thyroid	7.4E-09	5.2E-09	2.6E-09	1.6E-09	9.3E-10	7.6E-10
Uterus	7.4E-09	5.2E-09	2.6E-09	1.6E-09	9.3E-10	7.6E-10
Remainder	7.4E-09	5.2E-09	2.6E-09	1.6E-09	9.3E-10	7.6E-10
Effective Dose	7.6E-09	5.3E-09	2.7E-09	1.6E-09	9.5E-10	7.8E-10

GI-Tract	Gastrointestinal Tract
St	Stomach
SI	Small Intestine
ULI	Upper Large Intestine
LLI	Lower Large Intestine

References

Dziewiatkowski, D. D. (1945) Fate of ingested sulfate sulfur labeled with radioactive sulfur in the rat. *J. Biol. Chem.* **161**, 723–729.

Dziewiatkowski, D. D. (1949) On the utilization of exogenous sulfate by the rat in the formation of ethereal sulphates as indicated by the use of sodium sulphate labelled with radioactive sulphur. *J. Biol. Chem.* **178**, 389–393.

Dziewiatkowski, D. D. (1962) Sulphur. In: *Mineral Metabolism*, Ch. 28, Vol. II, Part B (C. L. Comar and F. Bronner, Eds). Academic Press, New York.

ICRP (1959) *Recommendations of the ICRP*, ICRP Publication 2. Pergamon Press, Oxford.

ICRP (1975) *Report of the Task Group on Reference Man*, ICRP Publication 23. Pergamon Press, Oxford.

ICRP (1980) *Limits for Intakes of Radionuclides by Workers: Part 2*, ICRP Publication 30. Pergamon Press, Oxford.

Maass, A. R., Larson, F. C. and Gordon, E. S. (1949) Distribution in normal tissues of radioactive sulfur fed as labeled methionine. *J. Biol. Chem.* **177**, 209–216.

Minski, M. and Cleary, J. J. (1966) The estimation of stable sulphur in human tissue. *Health Phys.* **12**, 837–842.

Minski, M. and Vennart, J. (1971) Maximum permissible intakes of ^{35}S-L-methionine and ^{35}S-sodium sulphate deduced from experiments on rats. *Health Phys.* **20**, 1–9.

Shohl, A. T. and Sato, A. (1923) A: acid–base metabolism II. Mineral metabolism. *J. Biol. Chem.* **58**, 257–266.

Vennart, J. and Ash, P. J. N. D. (1976) Derived limits for ^{35}S in food and air. *Health Phys.* **30**, 291–294.

Volwiler, W., Goldworthy, P. D., MacMartin, M. P., Wood, P. A., Mackay, I. R. and Fremont-Smith, K. (1955) Biosynthetic determination with radioactive sulphur of turnover rates of various plasma proteins in normal and cirrhotic men. *J. Clin. Invest.* **34**, 1126–1146.

Walser, M., Seldin, D. W. and Grollman, A. (1953) An evaluation of radiosulphate for the determination of the volume of extracellular fluid in man and dogs. *J. Clin. Invest.* **32**, 299–311.

Woodard, H. Q., Pentlow, K. S., Mayer, K., Laughlin, J. S. and Marcove, R. C. (1976) Distribution and retention of ^{35}S as sodium sulfate in man. *J. Nucl. Med.* **17**, 285–289.

Wright, J. B., Martin, P. G., Skellenger, M. C. and Moschette, D. S. (1960) Metabolic patterns in preadolescent children. III Sulfur balance on three levels of nitrogen uptake. *J. Nutr.* **72**, 314–316.

2. COBALT

Uptake to blood

(36) *Adults.* The absorption of cobalt (Co) is affected by the mass and the chemical form entering the body. Smith *et al.* (1972) measured ^{60}Co retention by whole-body counting following oral administration as the chloride. Values for absorption were about 3% (range 1.1–5%) with tracer quantities but increased to about 25% with added carrier (1.2 mg). For ^{60}Co given between meals absorption was less than when it was given after an overnight fast. Similar results were obtained by Paley and Sussman (1963) who measured 48-h cumulative excretion of ^{60}Co following oral administration as the chloride to 37 human subjects under various conditions. On the assumption that 48-h urinary excretion accounts for about 50% of activity entering the blood (Smith *et al.*, 1972; Letourneau and Jack, 1972) the results correspond to levels of absorption in fasting patients of about 8% when ^{60}Co was given carrier-free and 30% in the presence of carrier (12.5 mg Co). In normally fed subjects the results gave levels of absorption of 9% in the presence of carrier and 2% when carrier-free. Absorption decreased if ^{60}Co was attached to proteins. Conversely, Taylor (1962) found that absorption of ^{58}Co by rats decreased in response to increased mass administered, though this could have been due to the much higher mass concentrations used in rat experiments relative to body mass. Following administration of ^{58}Co as the glycinate with about 10 μg stable cobalt absorption was about 30%. With increasing mass of cobalt administered, 30–110 μg, absorption decreased to about 10%.

(37) Nishimura *et al.* (1976) administered ^{60}Co chloride and ^{58}Co vitamin B_{12} to rats and compared whole-body retention after oral and intravenous administration. Absorption was then estimated to be 3% for cobalt in its inorganic form and 70% for the vitamin-incorporated element. If fish or marine algae are maintained in sea water containing radioactive ^{60}Co before being fed to rats, fractional absorption of the element is much higher than that of inorganic cobalt. Thus Inaba *et al.* (1982) reported tissue retention in rats at 2 days after administration of ^{60}Co incorporated in fish as 30% of the activity administered, whereas a value of 3% was obtained after administration of $^{60}CoCl_2$. For ^{60}Co in crab meat, absorption by human volunteers was estimated to be about 5% (Ramsden, 1991).

(38) The available data indicate that absorption of cobalt is very variable. In *ICRP Publication 30* (ICRP, 1979) an f_1 value of 0.3 is recommended for organically complexed cobalt compounds, and for all inorganic compounds of the element, excluding oxides and hydroxides. The NEA Expert Group on gastrointestinal absorption considered an f_1 value of 0.3 to be a reasonable value for the absorption of cobalt by adult members of the general public (NEA/OECD, 1988). The range of results reported in human studies suggests that in most cases of normal adults the fractional absorption of trace quantities of cobalt is less than 0.1. An f_1 of 0.1 is therefore adopted here for adults.

(39) *Children.* The absorption of both ^{60}Co chloride and ^{60}Co vitamin B_{12} following oral administration can be much higher in young than in adult rats (Nishimura *et al.*, 1976). At 3 days after administration as the chloride, retention by 7-day-old rats (10%) was approximately 30 times the adult value (0.3%). Retention of the element incorporated into the vitamin B_{12} ($\cong$ 100%) was almost double the adult value (55%). A study of cobalt balance in girls aged between 6 and 10 years (Engel *et al.*, 1967) gave an

estimated absorption of 0.57. This relatively high cobalt absorption could be associated with the stimulation of iron absorption during periods of rapid growth. However, balance studies such as this are difficult to interpret and do not provide a satisfactory basis for determining f_1 values. As there is evidence that absorption of cobalt may be increased in childhood an f_1 of 0.3 is adopted for children from 1 to 15 years.

(40) In line with the value proposed by NEA/OECD (1988), an f_1 value for cobalt of 0.6 is adopted here as an average for the first year of life.

Distribution and retention

(41) *Adults.* Following the entry of cobalt into the blood a large fraction is rapidly excreted in urine. Thus by about 21 days after intravenous injection of ^{57}Co the total body retention was for hamsters, 2.0%; rats, 1.9%; guinea pigs, 3.5% (Collier *et al.*, 1989) and for mice, 1.1% (Talbot and Morgan, 1989). Retention at 45 days was less than 1% for baboons (André *et al.*, 1989) and 1.3% for Maccaca monkeys (Thomas *et al.*, 1976).

(42) The distribution and subsequent retention of ^{60}Co in adult rats following the intraperitoneal injection of the chloride was reported by Hollins and McCullough (1971). Between 1 and 72 days after injection the relative concentrations of cobalt in most tissues did not change significantly. In the period between 10 and 72 days after injection, however, the concentration in the liver, kidneys and blood decreased relative to other tissues, while that in the bone increased. From 10 to 72 days the liver retained about 25% of the whole body activity, the skeleton 24%, the muscle 21%, the kidneys 8% and the gastrointestinal tract 11%. By 386 days after injection, however, the skeleton contained 65% of the total body activity. In rats, following a single intravenous injection of the chloride or chronic administration in drinking water the liver and kidneys had highest concentrations in the short term and bone in the long term (Thomas *et al.*, 1976). Thomas *et al.* (1976) have also reported biological half-times for the long-term component of ^{60}Co retention of 495 days (0.9% of the administered activity), 309 days (0.7%), 183 days (3%) and 180 days (4%) for mice, rats, monkeys and dogs, respectively.

(43) Some information on the retention of ^{60}Co is available for human subjects. Data from whole-body counting and blood sampling of 11 humans after intravenous injection of ^{60}Co chloride showed that the level of ^{60}Co in plasma declined rapidly following administration. This resulted from uptake by tissues and renal excretion. An average of 22% (range 14–30%) of the administered activity was excreted in the urine of six subjects within 24 h (Smith *et al.*, 1972). Taylor (1959) measured the urinary excretion of ^{58}Co glycinate during the first 24 h following intravenous injection in 13 normal subjects: the average excretion was 34% with individual values ranging from 23% to 47% of the injected activity. The amounts of ^{60}Co excreted in urine and faeces by six subjects during periods ranging from 2 to 8 days after intravenous injection showed that the greater proportion of cobalt was eliminated by the kidneys, 22% of the activity on average appearing in the first day's urine. The faecal contribution to the total ^{60}Co excreted by each subject during the periods observed ranged from 5% to 30% with an average of about 15%, giving a mean urinary to faecal excretion ratio of about 6:1. Whole-body retention in two subjects could be described by the sum of four exponential components with half-times of 0.54 days (44% of activity administered), 5.6 days (32%), 63 days (13%) and 800 days (11%). External measurements suggested the liver retained 20% of the total body content. Although precise measurements could

not be made, the concentration of ^{60}Co in the liver appeared to be several times higher than in the body as a whole, even after 1000 days. The exact time course of retention in the tissue could not be established with any precision. These data were used as the basis for the retention functions for cobalt given in *ICRP Publication 30* (ICRP, 1979). This model assumes that of the cobalt entering the transfer compartment a fraction 0.5 goes directly to excretion, 0.05 to the liver and the remaining 0.45 is assumed to be uniformly distributed amongst all the other organs and tissues of the body. Fractions 0.6, 0.2 and 0.2 of the cobalt translocated from the transfer compartment to tissues are assumed to be retained with biological half-times of 6, 60 and 800 days, respectively. Cobalt in the transfer compartment is taken to be retained with a half-time of 0.5 days. This function is similar to that derived by Letourneau and Jack (1972) based on whole-body counting for 16 human subjects followed for a period of 1 year. This study gave whole-body retention parameters of 0.25 days (36% of the activity administered), 1.8 days (24%), 8 days (19%), 47 days (13%) and 608 days (9%).

(44) Although the data suggest that liver may not always retain the highest concentration of cobalt the available information is inadequate for making any more specific recommendations. For this report the retention parameters given by ICRP (1979) are therefore adopted. A urinary to faecal excretion ratio of 6:1 is assumed for cobalt that has entered the transfer compartment.

(45) *Children.* Nishimura *et al.* (1976) concluded that the retention of cobalt by rats was independent of age. They estimated that the increased whole-body retention of ^{60}Co chloride and ^{58}Co vitamin B_{12} by juvenile rats as compared with adults was due entirely to increased intestinal absorption. On the basis of these observations the retention parameters adopted for adults are also assumed for infants and children.

Dose coefficients

(46) Dose coefficients derived from the biokinetic data summarized in Table 2.-1 are given in Tables 2.-2 to 2.-4. Dose coefficients for cobalt vitamin B_{12} are given in *ICRP Publication 53* (ICRP, 1987).

Table 2.-1. Biokinetic data for cobalt

		Distribution (%)			Biological half-time (days)			
					Liver and other tissues			
Age	f_1	Other tissues	Liver	Prompt excretion	Comp. A (60%)	Comp. B (20%)	Comp. C (20%)	Transfer comp.
3 months	0.6	45	5	50	6	60	800	0.5
1 year	0.3	45	5	50	6	60	800	0.5
5 years	0.3	45	5	50	6	60	800	0.5
10 years	0.3	45	5	50	6	60	800	0.5
15 years	0.3	45	5	50	6	60	800	0.5
Adult	0.1	45	5	50	6	60	800	0.5

A urinary to faecal excretion ratio of 6:1 is assumed for cobalt that has entered the transfer compartment.

Table 2.-2

Ingestion Dose Coefficients (Equivalent Dose (Sv/Bq) to Age 70 Years) for Co-57 (T1/2 = 270.9 d)

Age at intake	3 Months	1 Year	5 Years	10 Years	15 Years	Adult
Adrenals	2.3E-09	9.5E-10	5.7E-10	3.8E-10	2.6E-10	7.8E-11
Bladder Wall	2.3E-09	1.0E-09	6.2E-10	4.4E-10	3.1E-10	1.2E-10
Bone Surfaces	3.0E-09	1.3E-09	7.5E-10	5.0E-10	3.5E-10	1.2E-10
Brain	1.8E-09	7.0E-10	4.1E-10	2.5E-10	1.6E-10	4.6E-11
Breast	1.5E-09	5.9E-10	3.3E-10	2.1E-10	1.4E-10	4.0E-11
GI-Tract						
St Wall	2.6E-09	1.2E-09	6.6E-10	4.4E-10	2.9E-10	1.2E-10
SI Wall	2.8E-09	1.6E-09	9.4E-10	6.2E-10	4.0E-10	2.4E-10
ULI Wall	4.8E-09	3.7E-09	2.0E-09	1.2E-09	7.6E-10	5.9E-10
LLI Wall	9.2E-09	8.1E-09	4.2E-09	2.6E-09	1.6E-09	1.4E-09
Kidneys	2.2E-09	9.0E-10	5.4E-10	3.6E-10	2.4E-10	8.0E-11
Liver	5.0E-09	2.0E-09	1.2E-09	8.3E-10	5.7E-10	1.7E-10
Lungs	2.0E-09	8.0E-10	4.7E-10	3.0E-10	2.1E-10	5.9E-11
Muscle	1.9E-09	7.7E-10	4.4E-10	2.8E-10	1.9E-10	6.5E-11
Ovaries	2.6E-09	1.3E-09	8.1E-10	5.4E-10	3.6E-10	2.0E-10
Pancreas	2.5E-09	1.0E-09	6.2E-10	4.1E-10	2.7E-10	8.9E-11
Red Marrow	1.8E-09	7.5E-10	4.5E-10	3.1E-10	2.1E-10	7.7E-11
Skin	1.5E-09	5.7E-10	3.2E-10	2.0E-10	1.3E-10	4.1E-11
Spleen	2.1E-09	8.7E-10	5.1E-10	3.3E-10	2.2E-10	7.0E-11
Testes	1.8E-09	7.4E-10	4.2E-10	2.7E-10	1.8E-10	5.9E-11
Thymus	1.9E-09	7.5E-10	4.3E-10	2.8E-10	1.9E-10	5.3E-11
Thyroid	2.1E-09	8.2E-10	4.8E-10	3.0E-10	1.9E-10	5.4E-11
Uterus	2.4E-09	1.1E-09	6.8E-10	4.5E-10	2.9E-10	1.3E-10
Remainder	1.9E-09	7.7E-10	4.5E-10	2.9E-10	1.9E-10	6.9E-11
Effective Dose	2.9E-09	1.6E-09	8.9E-10	5.8E-10	3.7E-10	2.1E-10

GI-Tract Gastrointestinal Tract
St Stomach
SI Small Intestine
ULI Upper Large Intestine
LLI Lower Large Intestine

Table 2.-3

Ingestion Dose Coefficients (Equivalent Dose (Sv/Bq) to Age 70 Years) for Co-58 (T1/2 = 70.80 d)

Age at intake	3 Months	1 Year	5 Years	10 Years	15 Years	Adult
Adrenals	6.1E-09	2.6E-09	1.5E-09	1.0E-09	6.9E-10	2.3E-10
Bladder Wall	5.7E-09	3.0E-09	2.0E-09	1.3E-09	8.3E-10	4.7E-10
Bone Surfaces	5.3E-09	2.2E-09	1.3E-09	8.3E-10	5.7E-10	2.2E-10
Brain	4.3E-09	1.6E-09	9.3E-10	5.8E-10	3.6E-10	1.0E-10
Breast	3.6E-09	1.4E-09	7.9E-10	5.1E-10	3.6E-10	1.1E-10
GI-Tract						
St Wall	6.7E-09	3.3E-09	1.9E-09	1.2E-09	8.2E-10	4.5E-10
SI Wall	8.6E-09	5.7E-09	3.4E-09	2.2E-09	1.4E-09	1.1E-09
ULI Wall	1.3E-08	9.9E-09	5.6E-09	3.7E-09	2.3E-09	2.0E-09
LLI Wall	2.1E-08	1.9E-08	1.1E-08	6.8E-09	4.2E-09	4.0E-09
Kidneys	5.8E-09	2.5E-09	1.5E-09	1.0E-09	6.7E-10	2.8E-10
Liver	1.0E-08	4.3E-09	2.5E-09	1.8E-09	1.2E-09	4.0E-10
Lungs	4.8E-09	1.9E-09	1.1E-09	7.3E-10	5.0E-10	1.5E-10
Muscle	4.9E-09	2.2E-09	1.2E-09	8.0E-10	5.4E-10	2.3E-10
Ovaries	8.5E-09	5.8E-09	3.4E-09	2.3E-09	1.6E-09	1.1E-09
Pancreas	6.8E-09	3.0E-09	1.7E-09	1.1E-09	7.6E-10	2.9E-10
Red Marrow	4.4E-09	1.9E-09	1.3E-09	9.0E-10	6.4E-10	3.1E-10
Skin	3.5E-09	1.4E-09	7.9E-10	4.9E-10	3.3E-10	1.3E-10
Spleen	5.6E-09	2.4E-09	1.4E-09	9.1E-10	6.0E-10	2.3E-10
Testes	4.6E-09	2.1E-09	1.2E-09	7.8E-10	5.1E-10	2.2E-10
Thymus	4.8E-09	1.9E-09	1.1E-09	6.9E-10	4.7E-10	1.4E-10
Thyroid	5.3E-09	2.0E-09	1.2E-09	7.3E-10	4.6E-10	1.3E-10
Uterus	7.0E-09	4.0E-09	2.3E-09	1.5E-09	9.9E-10	5.8E-10
Remainder	4.8E-09	2.1E-09	1.2E-09	8.2E-10	5.5E-10	2.5E-10
Effective Dose	7.4E-09	4.5E-09	2.6E-09	1.7E-09	1.1E-09	7.5E-10

GI-Tract   Gastrointestinal Tract
St   Stomach
SI   Small Intestine
ULI   Upper Large Intestine
LLI   Lower Large Intestine

Table 2.-4

Ingestion Dose Coefficients (Equivalent Dose (Sv/Bq) to Age 70 Years) for Co-60 (T1/2 = 5.271 y)

Age at intake	3 Months	1 Year	5 Years	10 Years	15 Years	Adult
Adrenals	5.7E-08	2.5E-08	1.6E-08	1.1E-08	8.0E-09	2.5E-09
Bladder Wall	4.6E-08	2.2E-08	1.4E-08	8.6E-09	6.4E-09	2.6E-09
Bone Surfaces	4.8E-08	2.1E-08	1.3E-08	8.8E-09	6.4E-09	2.1E-09
Brain	4.1E-08	1.7E-08	1.1E-08	6.7E-09	4.6E-09	1.4E-09
Breast	3.4E-08	1.4E-08	9.1E-09	6.1E-09	4.5E-09	1.4E-09
GI-Tract						
St Wall	5.1E-08	2.3E-08	1.5E-08	9.3E-09	7.0E-09	2.6E-09
SI Wall	6.0E-08	3.1E-08	2.0E-08	1.3E-08	9.0E-09	4.2E-09
ULI Wall	6.9E-08	4.1E-08	2.5E-08	1.6E-08	1.1E-08	6.6E-09
LLI Wall	8.7E-08	6.4E-08	3.7E-08	2.4E-08	1.5E-08	1.2E-08
Kidneys	5.3E-08	2.3E-08	1.5E-08	1.0E-08	7.2E-09	2.4E-09
Liver	9.9E-08	4.3E-08	2.8E-08	2.0E-08	1.5E-08	4.5E-09
Lungs	4.5E-08	1.9E-08	1.2E-08	8.3E-09	6.0E-09	1.8E-09
Muscle	4.4E-08	1.9E-08	1.2E-08	8.0E-09	5.7E-09	1.9E-09
Ovaries	5.9E-08	3.1E-08	1.9E-08	1.3E-08	9.1E-09	4.3E-09
Pancreas	6.1E-08	2.7E-08	1.7E-08	1.2E-08	8.2E-09	2.6E-09
Red Marrow	4.1E-08	1.8E-08	1.2E-08	8.6E-09	6.2E-09	2.2E-09
Skin	3.3E-08	1.4E-08	8.5E-09	5.5E-09	4.0E-09	1.3E-09
Spleen	5.0E-08	2.2E-08	1.4E-08	9.2E-09	6.5E-09	2.1E-09
Testes	4.1E-08	1.8E-08	1.2E-08	7.6E-09	5.5E-09	1.8E-09
Thymus	4.5E-08	1.9E-08	1.2E-08	8.1E-09	5.9E-09	1.7E-09
Thyroid	5.0E-08	2.1E-08	1.3E-08	8.4E-09	5.8E-09	1.7E-09
Uterus	5.6E-08	2.7E-08	1.7E-08	1.1E-08	7.9E-09	3.1E-09
Remainder	4.4E-08	1.9E-08	1.2E-08	8.0E-09	5.8E-09	2.0E-09
Effective Dose	5.4E-08	2.7E-08	1.7E-08	1.1E-08	7.9E-09	3.4E-09

GI-Tract Gastrointestinal Tract
St Stomach
SI Small Intestine
ULI Upper Large Intestine
LLI Lower Large Intestine

References

André, S., Métivier, H. and Masse, R. (1989) An interspecies comparison of the lung clearance of inhaled monodisperse cobalt oxide particles. Part 3: Lung clearance of inhaled cobalt oxide particles in baboons. *J. Aerosol Sci.* **20**, 205–217.

Collier, C. G., Bailey, M. R. and Hodgson, A. (1989) An interspecies comparison of the lung clearance of monodisperse cobalt oxide particles. Part 5: Lung clearance of inhaled cobalt oxide particles in hamsters, rats and guinea-pigs. *J. Aerosol Sci.* **20**, 233–247.

Engel, R. W., Price, N. O. and Miller, R. F. (1967) Copper, manganese, cobalt, and molybdenum balance in pre-adolescent girls. *J. Nutr.* **92**, 197–204.

Hollins, J. G. and McCullough, R. S. (1971) Radiation dosimetry of internal contamination by inorganic compounds of cobalt: an analysis of cobalt metabolism in rats. *Health Phys.* **21**, 233–246.

ICRP (1979) *Limits for Intakes of Radionuclides by Workers: Part 1*, ICRP Publication 30. Pergamon Press, Oxford.

ICRP (1987) *Radiation Dose to Patients from Radiopharmaceuticals*, ICRP Publication 53. Pergamon Press, Oxford.

Inaba, J., Nishimura, Y., Kimura, K. and Ichikawa, R. (1982) Whole-body retention and tissue distribution of cobalt-60 in rats after oral administration of freshwater fish contaminated with cobalt-60. *Health Phys.* **43**, 247–250.

Letourneau, E. G. and Jack, G. C. (1972) The metabolism of cobalt by the normal human male: whole body retention and radiation dosimetry. *Health Phys.* **22**, 451–459.

NEA/OECD (1988) Committee on Radiation Protection and Public Health. *Report of an Expert Group on Gut Transfer Factors.* NEA/OECD Report, Paris.

Nishimura, Y., Inaba, J. and Ichikawa, R. (1976) Whole-body retention of cobalt-60 chloride and cobalt-58 cyanocobalamin in young and adult rats. *J. Radiat. Res.* **17**, 240–246.

Paley, K. R. and Sussmann, E. S. (1963) Absorption of radioactive cobaltous chloride in human subjects. *Metabolism* **12**, 975–982.

Ramsden, D. (1991) Private communication. AEE, Winfrith, Dorset, UK.

Smith, T., Edmonds, C. J. and Barnaby, C. F. (1972) Absorption and retention of cobalt in man by whole body counting. *Health Phys.* **22**, 359–367.

Talbot, R. J. and Morgan, A. (1989) An interspecies comparison of the lung clearance of inhaled monodisperse cobalt oxide particles. Part 8: Lung clearance of inhaled cobalt oxide particles in mice. *J. Aerosol Sci.* **20**, 261–265.

Taylor, D. M. (1959) Metabolism of some trace metals in animals and man. PhD Thesis, University of London.

Taylor, D. M. (1962) The absorption of cobalt from the gastro-intestinal tract of the rat. *Phys. Med. Biol.* **6**, 445–451.

Thomas, R. G., Furchner, J. E., London, J. E., Drake, G. A., Wilson, J. S. and Richmond, C. R. (1976) Comparative metabolism of radionuclides in mammals—X. Retention of tracer level cobalt in the mouse, rat, monkey and dog. *Health Phys.* **31**, 323–333.

3. NICKEL

Uptake to blood

(47) *Adults.* When nickel sulphate was administered to human volunteers, nickel (Ni) levels in urine (Menne *et al.*, 1978) and serum (Christensen *et al.*, 1979) suggested absorption of about 2%. A balance study of 10 volunteers by Horak and Sunderman (1973) suggested a comparable value of around 1%. However, other balance studies have produced values of absorption around 50% (Tipton *et al.*, 1966; Veterans Administration Hospital, 1975), but each was based on only two subjects. Reviews of the biokinetics of nickel by Schroeder *et al.* (1961), Linder (1978) and Sunderman (1988) suggest an f_1 value of about 0.05. In *Publication 30*, ICRP (1981) adopted an f_1 value of 0.05 for all compounds of the element. This value is also adopted here.

(48) *Children.* As there appears to be no information available concerning intestinal absorption of nickel in young animals and neonates, the general approach of the NEA Expert Group (NEA/OECD, 1988) of an increase of the f_1 value for the adult by a factor of 2 to 0.1 is adopted here for the 3-month-old infant. For children 1 year and older the f_1 value for the adult (0.05) is used.

Distribution and retention

(49) *Adults.* In mice, rats and rabbits, ^{63}Ni was preferentially deposited in the kidneys following parenteral administration, with higher than average concentrations also found in the pituitary gland, lungs, skin, adrenals and gonads (Smith and Hackley, 1968; Parker and Sunderman, 1974; Jacobsen *et al.*, 1978; Olsen and Jonsen, 1979). Data from Reference Man (ICRP, 1975) indicate, however, that the concentration of stable nickel is similar in all soft tissues, suggesting that the early uptake by the kidneys is rapidly excreted in the urine and that there is no long-term selective accumulation by this and other tissues. Nickel has not been found to have any particular affinity for mineralized tissues (Jacobsen *et al.*, 1978; Sunderman, 1988). In this report, nickel is therefore assumed to be distributed throughout all organs and tissues of the body other than the kidney, where an initial increased uptake is assumed. In human volunteers after oral intake of nickel sulphate the half-time of clearance of nickel from the plasma, via urinary excretion, was approximately 11 h (Christensen and Lagesson, 1981). In common with *ICRP Publication 30* (ICRP, 1981) it is assumed here that of nickel leaving the transfer compartment, a fraction, 0.02, is accumulated by the kidneys and retained with a biological half-time of 0.2 days. A second fraction, 0.68, is assumed here to go directly to urinary excretion. The remainder of nickel leaving the transfer compartment (0.30) is assumed to be distributed throughout all organs and tissues of the body including the kidneys and to be retained with a half-time of 1200 days. These retention parameters are reasonably consistent with a daily intake of stable nickel in food and drink of 0.4 mg and a body content of 10.0 mg (ICRP, 1975).

(50) Studies in rats by Sunderman and Selin (1968) and Smith and Hackley (1968) have shown that about 5–10% of injected nickel excreted from the body is lost in the faeces. A urinary to faecal excretion ratio of 20:1 is therefore assumed here for nickel that has entered the transfer compartment.

(51) *Children.* No information appears to be available on the biokinetics of nickel in early infancy (Casey and Neville, 1987). The biokinetic data for the adult have therefore been adopted here for infants and children.

Dose coefficients

(52) Dose coefficients derived from the biokinetic data in Table 3.-1 are given in Tables 3.-2 and 3.-3.

Table 3.-1. Biokinetic data for nickel

		Distribution (%)			Biological half-time (days)	
Age	f_1	Prompt excretion	Kidneys	All organs*	Kidneys	All organs†
3 months	0.1	68	2	30	0.2	1200
1 year	0.05	68	2	30	0.2	1200
5 years	0.05	68	2	30	0.2	1200
10 years	0.05	68	2	30	0.2	1200
15 years	0.05	68	2	30	0.2	1200
Adult	0.05	68	2	30	0.2	1200

*Including kidneys.

†Includes long-term component in the kidneys.

A urinary to faecal excretion ratio of 20:1 is assumed for nickel that has entered the transfer compartment.

Table 3.-2

Ingestion Dose Coefficients (Equivalent Dose (Sv/Bq) to Age 70 Years) for Ni-59 (T1/2 = 7.5E4 y)

Age at intake	3 Months	1 Year	5 Years	10 Years	15 Years	Adult
Adrenals	3.4E-10	1.5E-10	8.9E-11	5.5E-11	4.0E-11	3.6E-11
Bladder Wall	3.5E-10	1.5E-10	9.2E-11	5.8E-11	4.2E-11	3.8E-11
Bone Surfaces	3.4E-10	1.5E-10	8.9E-11	5.5E-11	4.0E-11	3.6E-11
Brain	3.4E-10	1.5E-10	8.9E-11	5.5E-11	4.0E-11	3.6E-11
Breast	3.4E-10	1.5E-10	8.9E-11	5.5E-11	4.0E-11	3.6E-11
GI-Tract						
St Wall	4.5E-10	2.0E-10	1.2E-10	7.1E-11	5.0E-11	4.5E-11
SI Wall	5.6E-10	2.9E-10	1.6E-10	9.8E-11	6.3E-11	5.6E-11
ULI Wall	1.6E-09	1.0E-09	5.3E-10	3.2E-10	1.8E-10	1.5E-10
LLI Wall	4.0E-09	2.7E-09	1.3E-09	8.0E-10	4.6E-10	3.8E-10
Kidneys	3.4E-10	1.5E-10	8.9E-11	5.5E-11	4.0E-11	3.6E-11
Liver	3.4E-10	1.5E-10	8.9E-11	5.5E-11	4.0E-11	3.6E-11
Lungs	3.4E-10	1.5E-10	8.9E-11	5.5E-11	4.0E-11	3.6E-11
Muscle	3.4E-10	1.5E-10	8.9E-11	5.5E-11	4.0E-11	3.6E-11
Ovaries	3.4E-10	1.5E-10	8.9E-11	5.5E-11	4.0E-11	3.6E-11
Pancreas	3.4E-10	1.5E-10	8.9E-11	5.5E-11	4.0E-11	3.6E-11
Red Marrow	3.4E-10	1.5E-10	8.9E-11	5.5E-11	4.0E-11	3.6E-11
Skin	3.4E-10	1.5E-10	8.9E-11	5.5E-11	4.0E-11	3.6E-11
Spleen	3.4E-10	1.5E-10	8.9E-11	5.5E-11	4.0E-11	3.6E-11
Testes	3.4E-10	1.5E-10	8.9E-11	5.5E-11	4.0E-11	3.6E-11
Thymus	3.4E-10	1.5E-10	8.9E-11	5.5E-11	4.0E-11	3.6E-11
Thyroid	3.4E-10	1.5E-10	8.9E-11	5.5E-11	4.0E-11	3.6E-11
Uterus	3.4E-10	1.5E-10	8.9E-11	5.5E-11	4.0E-11	3.6E-11
Remainder	3.5E-10	1.5E-10	9.1E-11	5.6E-11	4.0E-11	3.7E-11
Effective Dose	6.3E-10	3.4E-10	1.9E-10	1.1E-10	7.3E-11	6.3E-11

GI-Tract Gastrointestinal Tract
St Stomach
SI Small Intestine
ULI Upper Large Intestine
LLI Lower Large Intestine

Table 3.-3

Ingestion Dose Coefficients (Equivalent Dose (Sv/Bq) to Age 70 Years) for Ni-63 (T1/2 = 96 y)

Age at intake	3 Months	1 Year	5 Years	10 Years	15 Years	Adult
Adrenals	8.2E-10	3.5E-10	2.1E-10	1.3E-10	9.4E-11	8.7E-11
Bladder Wall	8.4E-10	3.6E-10	2.2E-10	1.4E-10	9.9E-11	9.0E-11
Bone Surfaces	8.2E-10	3.5E-10	2.1E-10	1.3E-10	9.4E-11	8.7E-11
Brain	8.2E-10	3.5E-10	2.1E-10	1.3E-10	9.4E-11	8.7E-11
Breast	8.2E-10	3.5E-10	2.1E-10	1.3E-10	9.4E-11	8.7E-11
GI-Tract						
St Wall	1.1E-09	4.9E-10	2.8E-10	1.7E-10	1.2E-10	1.1E-10
SI Wall	1.4E-09	7.1E-10	3.9E-10	2.4E-10	1.5E-10	1.3E-10
ULI Wall	4.0E-09	2.5E-09	1.3E-09	7.8E-10	4.5E-10	3.7E-10
LLI Wall	9.9E-09	6.5E-09	3.3E-09	2.0E-09	1.1E-09	9.2E-10
Kidneys	8.3E-10	3.5E-10	2.1E-10	1.3E-10	9.5E-11	8.7E-11
Liver	8.2E-10	3.5E-10	2.1E-10	1.3E-10	9.4E-11	8.7E-11
Lungs	8.2E-10	3.5E-10	2.1E-10	1.3E-10	9.4E-11	8.7E-11
Muscle	8.2E-10	3.5E-10	2.1E-10	1.3E-10	9.4E-11	8.7E-11
Ovaries	8.2E-10	3.5E-10	2.1E-10	1.3E-10	9.4E-11	8.7E-11
Pancreas	8.2E-10	3.5E-10	2.1E-10	1.3E-10	9.4E-11	8.7E-11
Red Marrow	8.2E-10	3.5E-10	2.1E-10	1.3E-10	9.4E-11	8.7E-11
Skin	8.2E-10	3.5E-10	2.1E-10	1.3E-10	9.4E-11	8.7E-11
Spleen	8.2E-10	3.5E-10	2.1E-10	1.3E-10	9.4E-11	8.7E-11
Testes	8.2E-10	3.5E-10	2.1E-10	1.3E-10	9.4E-11	8.7E-11
Thymus	8.2E-10	3.5E-10	2.1E-10	1.3E-10	9.4E-11	8.7E-11
Thyroid	8.2E-10	3.5E-10	2.1E-10	1.3E-10	9.4E-11	8.7E-11
Uterus	8.2E-10	3.5E-10	2.1E-10	1.3E-10	9.4E-11	8.7E-11
Remainder	8.4E-10	3.6E-10	2.2E-10	1.3E-10	9.6E-11	8.8E-11
Effective Dose	1.5E-09	8.4E-10	4.6E-10	2.8E-10	1.8E-10	1.5E-10

GI-Tract Gastrointestinal Tract
St Stomach
SI Small Intestine
ULI Upper Large Intestine
LLI Lower Large Intestine

References

Casey, C. E. and Neville, M. C. (1987) Studies in human lactation 3: molybdenum and nickel in human milk during the first month of lactation. *Am. J. Clin. Nutr.* **45**, 921–926.

Christensen, O. B. and Lagesson, V. (1981) Nickel concentration of blood and urine after oral administration. *Ann. Clin. Lab. Sci.* **11**, 119–125.

Christensen, O. B., Moller, H., Andrasko, L. and Lagesson, V. (1979) Nickel concentration of blood, urine and sweat after oral administration. *Contact Derm.* **5**, 312–316.

Horak, E. and Sunderman, F. W. Jr. (1973) Fecal nickel excretion by healthy adults. *Clin. Chem.* **19**, 429–430.

ICRP (1975) *Report of the Task Group on Reference Man*, ICRP Publication 23. Pergamon Press, Oxford.

ICRP (1981) *Limits for Intakes of Radionuclides by Workers: Part 3*, ICRP Publication 30. Pergamon Press, Oxford.

Jacobsen, N., Alfheim, I. and Jonsen, J. (1978) Nickel and strontium distribution in some mouse tissues passage through placenta and mammary glands. *Res. Commun. Chem. Path. Pharmacol.* **20**, 571–584.

Linder, R. C. (1978) Functions and metabolism for trace elements. In: *Perinatal Physiology*, 2nd edn (U. Stave, Ed.). Plenum Press, New York.

Menne, T., Mikkelsen, H. and Solgard, P. (1978) Nickel excretion in urine after oral administration. *Contact Derm.* **4**, 106–108.

NEA/OECD (1988) Committee on Radiation Protection and Public Health. *Report of an Expert Group on Gut Transfer Factors.* NEA/OECD Report, Paris.

Olsen, I. and Jonsen, J. (1979) Whole-body autoradiography of ^{63}Ni in mice throughout gestation. *Toxicology* **12**, 165–172.

Parker, K. and Sunderman, F. W. Jr. (1974) Distribution of ^{63}Ni in rabbit tissues following intravenous injection of $^{63}NiCl_2$. *Res. Commun. Chem. Path. Pharmacol.* **7**, 755–762.

Schroeder, H., Balassa, J. and Tipton, I. (1961) Abnormal trace metals in man, nickel. *J. Chron. Dis.* **15**, 51–65.

Smith, J. C. and Hackley, B. (1968) Distribution and excretion of nickel-63 administered intravenously to rats. *J. Nutr.* **95**, 541–546.

Sunderman, F. W. Jr. (1988) Nickel. In: *Handbook on Toxicity of Inorganic Compounds* (H. Seiler and H. Sigel, Eds), pp. 453–468. Marcel Dekker, New York.

Sunderman, F. W. Jr. and Selin, C. E. (1968) The metabolism of nickel-63 carbonyl. *Toxicol. Appl. Pharmacol.* **12**, 207–218.

Tipton, I., Stewart, P. and Martin, P. (1966) Trace elements in diet and excreta. *Health Phys.* **12**, 1683–1689.

Veterans Administration Hospital, Hines, IL (1975) Metabolism of ^{90}Sr and of other elements in man. 1 July 1974–30 June 1975. COO-1231-104.

4. ZINC

Uptake to blood

(53) *Adults.* Tracer studies in which either stable or radioactive zinc (Zn) isotopes were incorporated in meals fed to normal adult volunteers have produced individual absorption values ranging up to about 60%, with a value of about 30% being typical (Solomons and Cousins, 1984; Sandstrom *et al.*, 1987, 1989). Studies in which zinc was administered in solution before breakfast generally gave greater values, from about 30% to 90% with 60% being typical (Molokhia *et al.*, 1980; Aamodt *et al.*, 1981; Istfan *et al.*, 1983; Mills *et al.*, 1983; Valberg *et al.*, 1986). Studies in which human volunteers have ingested crab meat containing ^{65}Zn gave f_1 values of about 0.15 (Ramsden, 1991), and an average value of 0.35 (range 0.25–0.45) was obtained in seven volunteers for ^{65}Zn in fish (Honstead and Brady, 1967). It has been suggested that some foods may enhance dietary zinc uptake (Solomons *et al.*, 1982), while bran and phytate reduce it (Turnland and King, 1983). Differences in dietary zinc levels also influence absorption (Furchner and Richmond, 1962). In *ICRP Publication 30* (ICRP, 1980) an f_1 value of 0.5 was adopted for all compounds of Zn. This value is also adopted here although it may be an overestimate for zinc in some foods.

(54) *Children.* Sullivan *et al.* (1984) showed an increase by a factor 4 in gastrointestinal absorption and tissue retention (67%) in 2-day-old neonatal rats compared with adult rats (16%). Ballou and Thompson (1961) showed that absorption fell abruptly after weaning in rats. Studies by Sherif *et al.* (1981) with rats showed a progressive decrease of gastrointestinal absorption with age from the weaning period (17–20 days) to 80 days. Ghishan *et al.* (1982) observed 90% absorption in suckling rats compared with 45% in adolescent rats given $^{65}ZnCl_2$ solution. Similar results have been reported for mice by He Liusheng *et al.* (1989). An f_1 value of 1 is adopted in this report for 3-month-old infants following the general approach described in paragraph (11) of the Introduction. For children of 1 year and older the f_1 value for the adult (0.5) is used here.

Distribution and retention

(55) *Adults.* The highest concentration of zinc is found in the prostate gland (Tipton and Cook, 1963), with values about three times those in other tissues. Spencer *et al.* (1965) reported that the whole-body retention of ^{65}Zn in humans could be described by the sum of two exponential components with biological half-times of 12 days (25%) and 320 days (75%). Two components of retention have also been reported by Richmond *et al.* (1962) in mice, rats, dogs and man. In four human subjects between 15% and 26% of the retained ^{65}Zn had a biological half-time between 4.5 and 26 days. The remainder had a half-time between 387 and 478 days. Hawkins *et al.* (1976) have reported biological half-times of about 20 days (15%) and 400 days (85%) for the two components of retention in eight patients with dermatological diseases. These results are fairly consistent. From the data of Richmond *et al.* (1962) it is clear that zinc is more tenaciously retained in the skeleton than it is in other tissues, and this is in agreement with the distribution of stable zinc given for Reference Man (ICRP, 1975). The gross distribution of zinc in the skeleton is similar to that of lead (Strehlow and Kneip, 1969).

Zinc is incorporated in mineral bone, where it is tenaciously retained. It is also found in the collagen matrix and in cartilage (NRC, 1979). Concentrations in cartilage vary with age, being lowest under 2–3 years of age and rising thereafter (NRC, 1979). In *ICRP Publication 30* (ICRP, 1980) 20% of zinc leaving the transfer compartment is assumed to be translocated to the skeleton. Of this, 19.5% is now assumed to be retained there with a biological half-time of 400 days, and 0.5% is assumed to be retained with a biological half-time of 10 000 days to account for retention in mineral bone. The remainder is assumed to be uniformly distributed throughout all other organs and tissues of the body and retained with biological half-times of 20 days (30%) and 400 days (70%). The retention parameters recommended by ICRP (1980) are therefore adopted in this report for the calculation of dose coefficients for adults in the population.

(56) Studies in humans (Craig and Siegel, 1960; Spencer *et al.*, 1965) have indicated that the total excretion of ^{65}Zn after intravenous injection as the chloride was quite low, averaging less than 10% in the first week in a total of 13 subjects. Excretion occurred predominantly in the faeces but the urinary to faecal excretion ratio varied quite widely in individual subjects, ranging from 5:1 to 1:11.2 with an average value of 1:4 for 10 subjects. A value of 1:4 for the urinary to faecal excretion ratio is therefore proposed here for zinc that has entered the transfer compartment.

(57) *Children.* The tissue distribution of absorbed ^{65}Zn in rats following ingestion as the chloride was similar in animals of different ages from neonates up to 200 days of age (Ballou and Thompson, 1961). In a further experiment with rats the skeleton to liver ratio of ingested ^{65}Zn was not afected by age (Sulliven *et al.*, 1984). Little information appears to be available concerning age-related changes in zinc biokinetics, although the long-term component of retention in mice seems to be largely unaffected by age at time of administration (He Liusheng *et al.*, 1989). Consequently, biokinetic parameters for the adult are used in this report for all ages for all soft tissues. For the skeleton, the half-times adopted for the long term component retained in bone volume are based on the turn-over rate of cortical bone as described in paragraph (7) of the Introduction.

Classification of isotopes for bone dosimetry

(58) It is assumed that ^{65}Zn (half-life 245 days) is uniformly distributed throughout the volume of mineral bone at all times after its deposition in the skeleton, and that shorter lived isotopes of zinc remain on bone surfaces.

Dose coefficients

(59) Dose coefficients derived from the biokinetic data in Table 4.-1 are given in Table 4.-2.

Table 4.-1. Biokinetic data for zinc

Age	f_1	Distribution (%)				Biological half-time (days)			
		Skeleton		Rest of body		Skeleton		Rest of body	
		Comp. A	Comp. B	Comp. A	Comp. B	Comp. A	Comp. B	Comp. A	Comp. B
3 months	1	19.5	0.5	24	56	400	100	20	400
1 year	0.5	19.5	0.5	24	56	400	300	20	400
5 years	0.5	19.5	0.5	24	56	400	500	20	400
10 years	0.5	19.5	0.5	24	56	400	1000	20	400
15 years	0.5	19.5	0.5	24	56	400	1500	20	400
Adult	0.5	19.5	0.5	24	56	400	10 000	20	400

A urinary to faecal excretion ratio of 1:4 is assumed for zinc that has entered the transfer compartment.

Table 4.-2

Ingestion Dose Coefficients (Equivalent Dose (Sv/Bq) to Age 70 Years) for Zn-65 (T1/2 = 243.9 d)

Age at intake	3 Months	1 Year	5 Years	10 Years	15 Years	Adult
Adrenals	4.3E-08	1.8E-08	1.1E-08	7.3E-09	5.1E-09	4.4E-09
Bladder Wall	3.1E-08	1.3E-08	9.2E-09	5.1E-09	3.6E-09	3.6E-09
Bone Surfaces	5.2E-08	2.2E-08	1.3E-08	8.9E-09	6.2E-09	5.5E-09
Brain	3.4E-08	1.4E-08	8.6E-09	5.5E-09	3.6E-09	3.2E-09
Breast	2.4E-08	1.0E-08	5.9E-09	3.9E-09	2.9E-09	2.5E-09
GI-Tract						
St Wall	3.2E-08	1.3E-08	8.7E-09	5.3E-09	4.0E-09	3.5E-09
SI Wall	3.9E-08	1.7E-08	1.1E-08	7.3E-09	5.0E-09	4.3E-09
ULI Wall	3.7E-08	1.7E-08	1.1E-08	7.1E-09	4.9E-09	4.3E-09
LLI Wall	4.0E-08	2.1E-08	1.3E-08	8.5E-09	5.7E-09	5.3E-09
Kidneys	3.9E-08	1.6E-08	9.8E-09	6.5E-09	4.5E-09	3.8E-09
Liver	3.7E-08	1.5E-08	9.2E-09	6.1E-09	4.2E-09	3.7E-09
Lungs	3.4E-08	1.4E-08	8.2E-09	5.5E-09	3.9E-09	3.3E-09
Muscle	3.4E-08	1.4E-08	8.4E-09	5.6E-09	3.9E-09	3.4E-09
Ovaries	4.1E-08	1.8E-08	1.1E-08	7.6E-09	5.3E-09	4.5E-09
Pancreas	4.1E-08	1.7E-08	1.0E-08	6.9E-09	4.8E-09	4.2E-09
Red Marrow	4.3E-08	1.8E-08	1.1E-08	7.9E-09	5.6E-09	4.8E-09
Skin	2.5E-08	1.0E-08	6.0E-09	3.8E-09	2.7E-09	2.4E-09
Spleen	3.7E-08	1.5E-08	9.3E-09	6.2E-09	4.3E-09	3.7E-09
Testes	3.0E-08	1.3E-08	7.7E-09	5.1E-09	3.6E-09	3.2E-09
Thymus	3.4E-08	1.4E-08	8.4E-09	5.6E-09	3.9E-09	3.3E-09
Thyroid	3.8E-08	1.6E-08	9.5E-09	6.2E-09	4.1E-09	3.6E-09
Uterus	3.9E-08	1.7E-08	1.0E-08	7.1E-09	4.9E-09	4.2E-09
Remainder	3.4E-08	1.4E-08	8.5E-09	5.6E-09	3.9E-09	3.4E-09
Effective Dose	3.6E-08	1.6E-08	9.8E-09	6.4E-09	4.5E-09	3.9E-09

GI-Tract	Gastrointestinal Tract
St	Stomach
SI	Small Intestine
ULI	Upper Large Intestine
LLI	Lower Large Intestine

References

Aamodt, R., Rumble, W., Johnston, G., Marley, E. and Henkin, R. (1981) Absorption of orally administered ^{65}Zn by normal human subjects. *Am. J. Clin. Nutr.* **34**, 2648–2652.

Ballou, J. E. and Thompson, R. C. (1961) Metabolism of zinc-65 in the rat, consideration of permissible exposure limits. *Health Phys.* **6**, 6–18.

Craig, F. A. and Siegel, E. (1960) Distribution in blood and excretion of Zn^{65} in man. *Proc. Soc. Exp. Biol. Med.* **104**, 391–394.

Furchner, J. E. and Richmond, C. R. (1962) Effect of dietary zinc on the absorption of orally administered Zn^{65}. *Health Phys.* **8**, 35–40.

Ghishan, F. K., Stroop, S. and Meneely, R. (1982) The effect of lactose on the intestinal absorption of calcium and zinc in the rat during maturation. *Pediatr. Res.* **16**, 566–568.

Hawkins, T., Marks, J. M., Plummser, U. M. and Greaves, M. W. (1976) Whole-body monitoring and other studies of zinc-65 metabolism in patients with dermatological disease. *Clin. Exp. Dermatol.* **1**, 243–252.

He Liusheng, Yan Xiaoshang and Wu Dechang (1989) Private communication to the Task Group.

Honstead, J. F. and Brady, D. N. (1967) The uptake and retention of ^{32}P and ^{65}Zn from the consumption of Columbia river fish. *Health Phys.* **13**, 455–463.

ICRP (1975) *Report of the Task Group on Reference Man*, ICRP Publication 23. Pergamon Press, Oxford.

ICRP (1980) *Limits for Intakes of Radionuclides by Workers: Part 2*, ICRP Publication 30. Pergamon Press, Oxford.

Istfan, N., Janghorbani, M. and Young, V. (1983) Absorption of stable zinc-70 in healthy young men in relation to zinc intake. *Am. J. Clin. Nutr.* **38**, 187–194.

Mills, P., Fell, G., Bessent, R., Nelson, L. and Russell, R. (1983) A study of zinc metabolism in alcoholic cirrhosis. *Clin. Sci.* **64**, 527–535.

Molokhia, M., Sturniolo, G., Shields, R. and Turnlund, L. (1980) A simple method for measuring zinc absorption in man using a short lived isotope. *Am. J. Clin. Nutr.* 881–886.

Newton, D. and Holmes, A. (1966) A case of accidental inhalation of zinc-65 and silver-110m. *Radiat. Res.* **29**, 403–412.

NRC (1979) *Committee on Medical and Biologic Effects of Environmental Pollutants, Subcommittee on Zinc.* National Academy of Sciences. University Park Press, Baltimore.

Ramsden, D. (1991) Private communication. AEE, Winfrith, Dorset, UK.

Richmond, C. R., Furchner, J. E., Trafton, G. A. and Langham, W. H. (1962) Comparative metabolism of radionuclides in mammals—I: Uptake and retention of orally administered ^{65}Zn by four mammalian species. *Health Phys.* **8**, 481–489.

Sandstrom, B., Davidsson, L., Kivisto, B., Hasselblad, C. and Cedarblad, A. (1987) The effect of vegetables and beet fibre on the absorption of zinc in humans from composite meals. *Br. J. Nutr.* **58**, 49–57.

Sandstrom, B., Almgren, A., Kivisto, B. and Cedarblad, A. (1989) Effect of protein level and protein source on zinc absorption in humans. *J. Nutr.* **119**, 48–53.

Sherif, Y. F., Schwarz, F. J. and Kirchgessner, M. (1981) Zum Einfluß von Alter und Geschlecht auf die intestinale in vitro Absorption von Zink und Mangan. *Arch. Tierernährung* **31**(9), 597–607.

Solomons, N. W. and Cousins, R. J. (1984) Zinc. In: *Current Topics in Nutrition Disease, Vol. 12. Absorption and Malabsorption of Mineral Nutrients* (N. W. Solomons and I. H. Rosenberg, Eds). Alan R. Liss, New York.

Solomons, N. W., Janghorbani, M., Ting, B. *et al.* (1982) Bioavailability of zinc from a diet based on isolated soy protein application in young men of the stable isotope zinc-70. *J. Nutr.* **112**, 1809–1821.

Spencer, H., Rosoff, B., Feldstein, A., Cohn, S. and Gusmano, E. (1965) Metabolism of zinc-65 in man. *Radiat. Res.* **24**, 432–445.

Strehlow, C. D. and Kneip, T. J. (1969) The distribution of Pb and Zn in the human skeleton. *Ann. Ind. Hyg. Assoc. J.* **30**, 372–378.

Sullivan, M. F., Miller, B. M. and Goebel, J. C. (1984) Gastrointestinal absorption of metals (^{51}Cr, ^{65}Zn, ^{95m}Tc, ^{109}Cd, ^{113}Sn, ^{147}Pm, and ^{238}Pu) by rats and swine. *Environ. Res.* **35**, 439–453.

Tipton, I. H. and Cook, M. J. (1963) Trace elements in human tissue. Part II. Adult subjects from the United States. *Health Phys.* **9**, 103–145.

Turnland, J. and King, J. (1983) Effect of cellulose and phytate on zinc absorption in man. *Fed. Proc.* **42**, 822.

Valberg, L., Flanagan, P. and Bondy, D. (1986) Zinc absorption in inflammatory bowel disease. *Digest. Dis. Sci.* **31**, 724–731.

5. STRONTIUM

Uptake to blood

(60) *Adults.* Human data on the absorption of dietary strontium (Sr) and of soluble forms of the element give values ranging from about 15% to about 45% (Snyder *et al.*, 1964; Fujita, 1965; Carr, 1967; Shimmins *et al.*, 1967; Spencer *et al.*, 1967; ICRP, 1973). Results for other animal species are generally similar (Coughtrey and Thorne, 1983). A number of factors have been found to increase absorption, including fasting and low dietary levels of calcium, magnesium and phosphorus. Milk diets and vitamin D may also increase absorption. Spencer *et al.* (1972) showed that overnight fasting of human volunteers resulted in an increase in strontium absorption from about 25% to about 55%. A decrease in the calcium content of the diet from 30–40 to 0–10 mg day^{-1} kg^{-1} increased strontium uptake in human volunteers from an average of 20% to 40% (Shimmins *et al.*, 1967). On the basis of the available information, ICRP (1979) assumed an f_1 of 0.3 for adults for all soluble forms of strontium. This value was recently endorsed by an Expert Group of the NEA/OECD (1988) and is adopted in this report for adults with a normal dietary intake of 1 g day^{-1}.

(61) *Children.* Results obtained by Widdowson *et al.* (1960) suggest that absorption of strontium in 7-day-old infants fed with cows' milk is greater than 73%. Bedford *et al.* (1960) reported that absorption in 5–15-year-old children was the same as in adults. However, studies on beagles and rats have shown that the period of increased absorption of strontium extends beyond the time of weaning. In the beagle, retention values for strontium at 3–9 days after ingestion were 20%, 15% and 8% in 48-, 80- and 140-day-old animals, respectively (Della Rosa *et al.*, 1965). The absorption of strontium in 35- and 75-day-old rats was estimated as 70–90% by Gran (1960) compared with 12% in 270-day-old rats. An Expert Group of the NEA/OECD (1988) has recently proposed an average f_1 of 0.6 for the first year of life. The value of 0.6 is adopted in this report for infants. For ages 1–15 years a value of 0.4 is used. The choice of this value is based on the consideration that there may be elevated absorption of strontium throughout the period of growth.

Distribution and retention

(62) Since the early 1950s a large quantity of data on the age-specific behaviour of strontium in humans has been published. Concentrations of fallout ^{90}Sr have been measured in thousands of human bones from persons of various ages. Also, numerous studies have dealt with the body's discrimination between strontium and calcium at different ages. Several age-dependent biokinetic models for strontium have been derived from these data (Kulp and Schulert, 1962; Bennett, 1972; Papworth and Vennart, 1984). Another age-dependent model for strontium was derived primarily from considerations of age-specific changes in skeletal calcium, together with information on age-specific discrimination between calcium and strontium by the human body (Leggett *et al.*, 1982, 1984). This model proved to be a reliable predictor of fallout data for ^{90}Sr and has the advantage for applications to other exposure scenarios that parameter values were derived to a large extent from basic physiological information on the human skeleton rather than strictly from empirical fits to ^{90}Sr data for chronic exposures.

Nevertheless, it was considered that the accuracy of this model for application to relatively short-lived strontium isotopes, including ^{85}Sr, could be improved by introducing further physiological considerations. The biokinetic model for strontium used in this report (Leggett, 1992) is a modification of the model of Leggett *et al.* (1982) that depicts exchange of material between blood, soft tissues, and the skeleton and explicitly addresses urinary and faecal excretion pathways. The conceptual framework for this model is the same as that used in this report for barium, radium, and lead. The model is described in Appendix A, and parameter values for strontium are given in Table A-3.

Dose coefficients

(63) Dose coefficients derived from the biokinetic data summarized in Table A-3 are given in Tables 5.-1 and 5.-2.

Table 5.-1

Ingestion Dose Coefficients (Equivalent Dose (Sv/Bq) to Age 70 Years) for Sr-89 (T1/2 = 50.5 d)*

Age at intake	3 Months	1 Year	5 Years	10 Years	15 Years	Adult
Adrenals	2.9E-09	1.7E-09	8.6E-10	4.5E-10	2.2E-10	2.0E-10
Bladder Wall	3.9E-09	2.6E-09	1.9E-09	1.1E-09	6.2E-10	6.7E-10
Bone Surfaces	1.6E-07	4.9E-08	3.1E-08	2.5E-08	2.2E-08	6.0E-09
Brain	2.9E-09	1.7E-09	8.6E-10	4.5E-10	2.2E-10	2.0E-10
Breast	2.9E-09	1.7E-09	8.6E-10	4.5E-10	2.2E-10	2.0E-10
GI-Tract						
St Wall	1.2E-08	6.4E-09	3.1E-09	1.7E-09	1.1E-09	8.8E-10
SI Wall	1.1E-08	9.3E-09	4.7E-09	2.7E-09	1.5E-09	1.4E-09
ULI Wall	5.6E-08	5.3E-08	2.6E-08	1.5E-08	8.2E-09	7.8E-09
LLI Wall	1.5E-07	1.4E-07	7.2E-08	4.2E-08	2.3E-08	2.2E-08
Kidneys	2.9E-09	1.7E-09	8.6E-10	4.5E-10	2.2E-10	2.0E-10
Liver	2.9E-09	1.7E-09	8.6E-10	4.5E-10	2.2E-10	2.0E-10
Lungs	2.9E-09	1.7E-09	8.6E-10	4.5E-10	2.2E-10	2.0E-10
Muscle	2.9E-09	1.7E-09	8.6E-10	4.5E-10	2.2E-10	2.0E-10
Ovaries	2.9E-09	1.7E-09	8.6E-10	4.5E-10	2.2E-10	2.0E-10
Pancreas	2.9E-09	1.7E-09	8.6E-10	4.5E-10	2.2E-10	2.0E-10
Red Marrow	1.6E-07	3.7E-08	1.7E-08	1.5E-08	1.4E-08	4.8E-09
Skin	2.9E-09	1.7E-09	8.6E-10	4.5E-10	2.2E-10	2.0E-10
Spleen	2.9E-09	1.7E-09	8.6E-10	4.5E-10	2.2E-10	2.0E-10
Testes	2.9E-09	1.7E-09	8.6E-10	4.5E-10	2.2E-10	2.0E-10
Thymus	2.9E-09	1.7E-09	8.6E-10	4.5E-10	2.2E-10	2.0E-10
Thyroid	2.9E-09	1.7E-09	8.6E-10	4.5E-10	2.2E-10	2.0E-10
Uterus	2.9E-09	1.7E-09	8.6E-10	4.5E-10	2.2E-10	2.0E-10
Remainder	3.1E-09	1.9E-09	9.6E-10	5.0E-10	2.4E-10	2.2E-10
Effective Dose	3.6E-08	1.8E-08	8.9E-09	5.8E-09	4.0E-09	2.6E-09

GI-Tract Gastrointestinal Tract
St Stomach
SI Small Intestine
ULI Upper Large Intestine
LLI Lower Large Intestine

* In the biokinetic model for Sr parameter values for the adult apply to ages > 25 y. For radioisotopes of this element the dose coefficients for the adult are based on the 50-y integrated doses following an acute intake at age 25 y.

Table 5.-2

Ingestion Dose Coefficients (Equivalent Dose (Sv/Bq) to Age 70 Years) for Sr-90 (T1/2 = 29.12 y)*

Age at intake	3 Months	1 Year	5 Years	10 Years	15 Years	Adult
Adrenals	1.2E-08	5.5E-09	2.9E-09	1.8E-09	1.1E-09	6.6E-10
Bladder Wall	1.6E-08	7.6E-09	5.0E-09	3.6E-09	2.7E-09	1.5E-09
Bone Surfaces	2.3E-06	7.3E-07	6.4E-07	1.1E-06	1.8E-06	4.1E-07
Brain	1.2E-08	5.5E-09	2.9E-09	1.8E-09	1.1E-09	6.6E-10
Breast	1.2E-08	5.5E-09	2.9E-09	1.8E-09	1.1E-09	6.6E-10
GI-Tract						
St Wall	1.5E-08	7.1E-09	3.7E-09	2.2E-09	1.4E-09	9.1E-10
SI Wall	1.5E-08	8.5E-09	4.4E-09	2.6E-09	1.6E-09	1.1E-09
ULI Wall	6.0E-08	4.3E-08	2.2E-08	1.3E-08	7.3E-09	5.9E-09
LLI Wall	1.9E-07	1.5E-07	7.5E-08	4.4E-08	2.5E-08	2.2E-08
Kidneys	1.2E-08	5.5E-09	2.9E-09	1.8E-09	1.1E-09	6.6E-10
Liver	1.2E-08	5.5E-09	2.9E-09	1.8E-09	1.1E-09	6.6E-10
Lungs	1.2E-08	5.5E-09	2.9E-09	1.8E-09	1.1E-09	6.6E-10
Muscle	1.2E-08	5.5E-09	2.9E-09	1.8E-09	1.1E-09	6.6E-10
Ovaries	1.2E-08	5.5E-09	2.9E-09	1.8E-09	1.1E-09	6.6E-10
Pancreas	1.2E-08	5.5E-09	2.9E-09	1.8E-09	1.1E-09	6.6E-10
Red Marrow	1.5E-06	4.2E-07	2.7E-07	3.7E-07	4.8E-07	1.8E-07
Skin	1.2E-08	5.5E-09	2.9E-09	1.8E-09	1.1E-09	6.6E-10
Spleen	1.2E-08	5.5E-09	2.9E-09	1.8E-09	1.1E-09	6.6E-10
Testes	1.2E-08	5.5E-09	2.9E-09	1.8E-09	1.1E-09	6.6E-10
Thymus	1.2E-08	5.5E-09	2.9E-09	1.8E-09	1.1E-09	6.6E-10
Thyroid	1.2E-08	5.5E-09	2.9E-09	1.8E-09	1.1E-09	6.6E-10
Uterus	1.2E-08	5.5E-09	2.9E-09	1.8E-09	1.1E-09	6.6E-10
Remainder	1.2E-08	5.5E-09	2.9E-09	1.8E-09	1.1E-09	6.7E-10
Effective Dose	2.3E-07	7.2E-08	4.7E-08	6.0E-08	7.9E-08	2.8E-08

GI-Tract Gastrointestinal Tract
St Stomach
SI Small Intestine
ULI Upper Large Intestine
LLI Lower Large Intestine

* In the biokinetic model for Sr parameter values for the adult apply to ages > 25 y. For radioisotopes of this element the dose coefficients for the adult are based on the 50-y integrated doses following an acute intake at age 25 y.

References

Bedford, J., Harrison, G. E., Raymond, W. H. A. and Sutton, A. (1960) The metabolism of strontium in children. *Br. Med. J.* **1**, 589–592.

Bennett, B. G. (1972) Fallout Sr-90 in diet and human bone. *Proceedings of the 2nd Int. Conf. on Strontium Metabolism* (J. M. A. Lenihan, Ed.), pp. 457–468.

Carr, T. E. I. (1967) An attempt to quantitate the short-term movement of strontium in human adult. In: *Strontium Metabolism* (J. M. A. Lenihan, J. F. Loutit and J. H. Martin, Eds), pp. 139–148. Academic Press, London.

Coughtrey, P. J. and Thorne, M. C. (1983) *Radionuclide Distribution and Transport in Terrestrial and Aquatic Ecosystems, Vol. I.* Balkema, Rotterdam.

Della Rosa, R. J., Goldman, A. C., Anderson, A. C., Mays, C. W. and Stover, B. J. (1965) Absorption and retention of ingested strontium and calcium in beagles. *Nature* **205**, 197–198.

Fujita, M. (1965) Absorption of strontium-90 in man. *Health Phys.* **11**, 47–50.

Gran, F. C. (1960) Studies on calcium and strontium-90 metabolism in rats. *Acta Physiol. Scand.* **48**, Suppl. No. 167.

ICRP (1973) *Alkaline Earth Metabolism in Adult Man*, ICRP Publication 20. Pergamon Press, Oxford.

ICRP (1979) *Limits for Intakes of Radionuclides by Workers: Part 1*, ICRP Publication 30. Pergamon Press, Oxford.

Kulp, J. L. and Schulert, A. R. (1962) Strontium-90 in man: V. *Science* **136**, 3516, 619.

Leggett, R. W. (1992) A generic age-specific biokinetic model for calcium-like elements. *Radiat. Prot. Dosim.* **41**, 183–198.

Leggett, R. W., Eckerman, K. F. and Williams, L. R. (1982) Strontium-90 in bone: a case study in age-dependent dosimetry modeling. *Health Phys.* **43**, 307–322.

Leggett, R. W., Eckerman, K. F., Dunning, D. E., Cristy, M., Crawford-Brown, D. J. and Williams, L. R. (1984) On estimating dose rates to organs as a function of age following internal exposure to radionuclides. Oak Ridge National Laboratory, Oak Ridge, ORNL/TM-8265.

NEA/OECD (1988) Committee on Radiation Protection and Public Health. *Report of an Expert Group on Gut Transfer Factors.* NEA/OECD Report, Paris.

Papworth, D. G. and Vennart, J. (1984) The uptake and turnover of Sr-90 in the human skeleton. *Phys. Med. Biol.* **29**, 1045–1061.

Shimmins, J., Smith, D. A., Nordin, B. E. C. and Burkinshaw, L. A. (1967) A comparison between calcium-45 and strontium-85 absorption, excretion and skeletal uptake. In: *Strontium Metabolism* (J. M. A. Lenihan, J. F. Loutit and J. H. Martin, Eds), pp. 149–159. Academic Press, London.

Snyder, W. S., Cook, M. J. and Ford, M. R. (1964) Estimates of (MPC) for occupational exposure to Sr-90, Sr-89 and Sr-85. *Health Phys.* **10**, 171–182.

Spencer, H., Lewin, I. and Samachson, J. (1967) Influence of dietary and hormonal factors on radiostrontium metabolism in man. In: *Strontium Metabolism* (J. M. A. Lenihan, J. F. Loutit and J. H. Martin, Eds), pp. 111–129. Academic Press, London.

Spencer, H., Kramer, L., Norris, C. and Samachson, J. (1972) Certain aspects of radiostrontrium metabolism in man. In: *Proceedings of the 2nd Int. Symp. on Strontium Metabolism* (J. M. A. Lenihan, Ed.), pp. 335–346.

Widdowson, E. M., Slater, E. J., Harrison, G. E. and Sutton, A. (1960) Absorption, excretion and retention of strontium by breast-fed and bottle-fed babies. *Lancet* **2**, 941–944.

6. MOLYBDENUM

Uptake to blood

(64) *Adults.* Molybdenum (Mo) is generally absorbed fairly readily from the gastrointestinal tract, even if it is administered in insoluble forms such as MoO_3 and $CaMoO_4$. However, MoS_2 is only poorly absorbed (ICRP, 1979). The gastrointestinal absorption of molybdenum is considered to be dependent on its concentration in the diet, the amounts of copper and sulphur present, and the age of the animal (Nederbragt, 1983). Experimental data show a wide variation in the fractional absorption of molybdenum, presumably owing to these factors. A value of 0.8 was recommended by ICRP (1979) for all compounds of molybdenum other than the sulphide, for which an f_1 value of 0.05 was given. In view of the reported variation of the f_1 value with diet a value of 1 has been adopted here following the general approach described in paragraph (10) of the Introduction.

(65) *Children.* Metabolic balance data were obtained for 36 pre-adolescent girls, aged 6–10 years, which showed that there is considerable variation in retention of molybdenum from various diets (Engel *et al.*, 1967). An f_1 value of 0.63–0.78 was derived from the data on excretion for this group of children (Coughtrey and Thorne, 1983). Following the general approach of paragraph (11) of the Introduction the same f_1 value (1) for infants and children as for adults has been adopted here.

Distribution and retention

(66) *Adults.* Information on the distribution of stable molybdenum in human tissues shows high concentrations in bone, liver and kidneys; the bone is also a site of long-term retention (Coughtrey and Thorne, 1983). When ^{99}Mo is intravenously administered as ammonium molybdate to mice, the liver shows the highest uptake with retention of about 26% of the administered activity at 1 h and about 21% at 24 h. The ^{99}Mo content of the kidney is relatively high, accounting for about 3.8% of the administered activity at 1 and 24 h (Rosoff and Spencer, 1973). When molybdenum was administered to rats as ammonium molybdate 74% was excreted within 3 h (Ando *et al.*, 1989). The tissue distribution was similar to that reported for mice. In domestic pigs exposed to radioactive fallout from a nuclear weapon test, significant amounts of ^{99}Mo were found in the liver and kidney. The high level of ^{99}Mo in the kidney can be attributed to renal excretion which rapidly removes molybdenum from the blood (Chertok and Lake, 1971).

(67) In four patients aged 53–78 years, the main route of excretion of ^{99}Mo administered carrier-free in ammonium hydroxide was urine with cumulative urinary excretion of about 12% (range 6.8–16%) in the first day and about 25% in 10 days. Only a small percentage (about 3%) was excreted in the faeces over this time (Rosoff and Spencer, 1964).

(68) On the basis of the human data the whole-body retention of molybdenum in humans has been described by two components of retention with half-times of 1 day (10%) and 50 days (90%) (ICRP, 1979). However, this retention function is not consistent with a daily intake of molybdenum of 0.034–0.34 mg and average soft tissue and skeleton contents of 15 and 300 mg, respectively (Coughtrey and Thorne, 1983). These data suggest that molybdenum is retained in the skeleton with a long half-time.

(69) In this report, for molybdenum entering the transfer compartment, 10% is assumed to be deposited in the skeleton and to be retained with a biological half-time of 10 000 days, consistent with the general approach outlined in paragraph (7) of the Introduction. The remaining activity is distributed to liver (25%), kidneys (5%) and all other tissues (60%). For molybdenum translocated to organs or tissues other than the skeleton, fractions of 0.1 and 0.9 are assumed to be retained with half-times of 1 and 50 days, respectively. A urinary to faecal excretion ratio of 8:1 is assumed for molybdenum that has entered the transfer compartment.

(70) *Children.* Limited data for rats show a shorter half-time retention in young animals compared with adults. These data do not provide a basis for age-dependent data for man and the adult parameters are adopted here for all ages.

Classification of isotopes for bone dosimetry

(71) It is assumed that molybdenum isotopes with half-lives of more than 15 days are uniformly distributed throughout the volume of mineral bone, and that all shorter lived isotopes of molybdenum remain on bone surfaces.

Dose coefficients

(72) Dose coefficients derived from the biokinetic data summarized in Table 6.-1 are given in Table 6.-2.

Table 6.-1. Biokinetic data for molybdenum

		Distribution (%)				Biological half-time (days)		
							Tissues other than skeleton	
Age	f_1	Skeleton	Liver	Kidney	Other tissues	Skeleton	Comp. A (10%)	Comp. B (90%)
3 months	1	10	25	5	60	10 000	1	50
1 year	1	10	25	5	60	10 000	1	50
5 years	1	10	25	5	60	10 000	1	50
10 years	1	10	25	5	60	10 000	1	50
15 years	1	10	25	5	60	10 000	1	50
Adult	1	10	25	5	60	10 000	1	50

A urinary to faecal excretion ratio of 8:1 is assumed for molybdenum that has entered the transfer compartment.

Table 6.-2

Ingestion Dose Coefficients (Equivalent Dose (Sv/Bq) to Age 70 Years) for Mo-99 (T1/2 = 66.0 h)

Age at intake	3 Months	1 Year	5 Years	10 Years	15 Years	Adult
Adrenals	2.9E-09	2.0E-09	1.0E-09	6.6E-10	4.1E-10	3.4E-10
Bladder Wall	2.9E-09	2.0E-09	1.2E-09	7.8E-10	5.5E-10	4.3E-10
Bone Surfaces	6.9E-09	5.2E-09	3.3E-09	1.9E-09	1.1E-09	1.0E-09
Brain	2.5E-09	1.6E-09	8.2E-10	4.9E-10	3.0E-10	2.4E-10
Breast	2.5E-09	1.6E-09	8.1E-10	4.9E-10	3.0E-10	2.4E-10
GI-Tract						
St Wall	8.6E-09	5.0E-09	2.5E-09	1.4E-09	9.5E-10	7.5E-10
SI Wall	2.9E-09	1.9E-09	9.9E-10	6.0E-10	3.6E-10	3.0E-10
ULI Wall	4.3E-09	2.9E-09	1.5E-09	8.8E-10	5.2E-10	4.2E-10
LLI Wall	6.2E-09	4.0E-09	2.0E-09	1.2E-09	7.1E-10	5.8E-10
Kidneys	2.4E-08	1.5E-08	8.2E-09	5.5E-09	3.9E-09	3.1E-09
Liver	2.4E-08	1.6E-08	8.4E-09	5.5E-09	3.6E-09	2.8E-09
Lungs	2.7E-09	1.8E-09	8.9E-10	5.5E-10	3.4E-10	2.8E-10
Muscle	2.6E-09	1.7E-09	8.5E-10	5.2E-10	3.2E-10	2.6E-10
Ovaries	2.7E-09	1.8E-09	9.1E-10	5.6E-10	3.4E-10	2.8E-10
Pancreas	3.0E-09	2.0E-09	1.0E-09	6.5E-10	4.0E-10	3.3E-10
Red Marrow	8.2E-09	5.0E-09	2.4E-09	1.4E-09	8.1E-10	6.1E-10
Skin	2.5E-09	1.6E-09	7.9E-10	4.8E-10	2.9E-10	2.4E-10
Spleen	2.7E-09	1.8E-09	9.3E-10	5.7E-10	3.5E-10	2.8E-10
Testes	2.5E-09	1.6E-09	8.1E-10	4.9E-10	3.0E-10	2.5E-10
Thymus	2.6E-09	1.7E-09	8.5E-10	5.2E-10	3.2E-10	2.6E-10
Thyroid	2.6E-09	1.7E-09	8.6E-10	5.2E-10	3.1E-10	2.5E-10
Uterus	2.7E-09	1.8E-09	9.0E-10	5.5E-10	3.3E-10	2.7E-10
Remainder	3.0E-09	1.9E-09	9.8E-10	5.9E-10	2.1E-09	1.7E-09
Effective Dose	5.5E-09	3.5E-09	1.8E-09	1.1E-09	7.6E-10	6.1E-10

GI-Tract Gastrointestinal Tract
St Stomach
SI Small Intestine
ULI Upper Large Intestine
LLI Lower Large Intestine

References

Ando, A., Ando, I., Hiraki, T. and Hisada, K. (1989) Relation between the location of elements in the Periodic Table and various organ-uptake rates. *Nucl. Med. Biol.* **16**, 57–80.

Chertok, R. J. and Lake, S. (1971) Biological availability of radionuclides. Produced by the Plowshare Event Schooner—I. Body distribution in domestic pigs exposed in the field. *Health Phys.* **20**, 317–324.

Coughtrey, P. J. and Thorne, M. C. (1983) *Radionuclide Distribution and Transport in Terrestrial and Aquatic Ecosystems, Vol. 2.* A. A. Balkema, Rotterdam.

Engel, R. W., Price, N. O. and Miller, R. F. (1967) Copper, manganese, cobalt, and molybdenum balance in pre-adolescent girls. *J. Nutr.* **92**, 197–204.

ICRP (1979) *Limits for Intakes of Radionuclides by Workers: Part 1,* ICRP Publication 30. Pergamon Press, Oxford.

Nederbragt, H. (1983) The influence of molybdenum on the copper metabolism of the rat at different Cu levels of the diet. *Br. J. Nutr.* **43**, 329.

Rosoff, B. and Spencer, H. (1964) Fate of molybdenum-99 in man. *Nature* **202**, 410–411.

Rosoff, B. and Spencer, H. (1973) The distribution and excretion of molybdenum-99 in mice. *Health Phys.* **25**, 173–175.

7. TECHNETIUM

Uptake to blood

(73) *Adults.* The results of Beasley *et al.* (1966) indicate that when technetium (Tc) was orally administered to humans in the form of pertechnetate ($^{95m}TcO_4$ and $^{96}TcO_4$) after an overnight fast, about 96% was absorbed from the gut. However, a marked variability in the absorption of technetium following an oral dose of pertechnetate has been observed between subjects and between separate studies in a single subject (Hays, 1972). In *ICRP Publication 30* (ICRP, 1980), an f_1 value of 0.8 is recommended for all inorganic compounds of the element.

(74) The fractional absorption of technetium from the gut is reduced to half of that following administration of pertechnetate, when soya-bean and animal tissue with incorporated technetium were fed to rats and guinea pigs (Sullivan *et al.*, 1977). Similar results in rats and sheep fed technetium either as pertechnetate or incorporated in maize have been obtained by Gerber *et al.* (1989). As technetium in food is less readily absorbed than the pertechnetate, an f_1 value of 0.5 is adopted here.

(75) *Children.* Technetium gavaged in the form of pertechnetate is well absorbed by both adult and neonatal rats (Sullivan *et al.*, 1984). However, there appear to be no additional data on the age dependence of the gastrointestinal absorption of technetium. Based on the general approach described in paragraph (11) of the Introduction a value of 1 is applied for 3-month-old infants. For children of 1 year and older the f_1 value for the adult (0.5) is used here.

Distribution and retention

(76) *Adults.* Beasley *et al.* (1966) reported that the whole-body retention of technetium in humans following intravenous administration as pertechnetate can be described by three components of retention with half-times of 1.6 days (77%), 3.7 days (19%) and 22 days (4%).

(77) In humans technetium administered intravenously as pertechnetate is concentrated in the thyroid, gastrointestinal tract and liver. Based on the information available, ICRP (1980) adopted a simplified model to describe the distribution and retention of technetium after entry into body fluids. Of technetium leaving the transfer compartment 0.04 is assumed to be translocated to the thyroid and retained with a half-time of 0.5 days. Further fractions, 0.1 and 0.03, are assumed to be translocated to the stomach wall and liver, respectively. The remaining fraction is assumed to be uniformly distributed throughout all organs and tissues of the body other than the thyroid, stomach wall and liver. Of technetium translocated to any organ or tissue of the body other than the thyroid, 0.75, 0.20 and 0.05 are assumed to be retained with biological half-times of 1.6, 3.7 and 22 days, respectively. For technetium the biological half-time in the transfer compartment is taken to be 0.02 days. No further data appear to have been published which suggest these biokinetic parameters need to be modified and they are therefore adopted here. Based on animal studies following administration by injection or ingestion of pertechnetate, approximately equal amounts of the element was excreted in urine and faeces (Archimbaud *et al.*, 1992). A urinary to faecal excretion ratio of 1:1 is therefore assumed for technetium that has entered the transfer compartment.

(78) *Children.* No information appears to be available for assessing the biokinetics of technetium in infants and children. The biokinetic data for the adult are therefore adopted here for infants and children.

Dose coefficients

(79) Dose coefficients derived from the biokinetic data summarized in Table 7.-1 are given in Tables 7.-2 and 7.-3.

Table 7.-1. Biokinetic data for technetium

		Distribution (%)				Biological half-time (days)				
							Tissues other than thyroid			
Age	f_1	Thyroid	Stomach wall	Liver	Other tissues	Thyroid	Comp. A (75%)	Comp. B (20%)	Comp. C (5%)	Transfer comp.
3 months	1	4	10	3	83	0.5	1.6	3.7	22	0.02
1 year	0.5	4	10	3	83	0.5	1.6	3.7	22	0.02
5 years	0.5	4	10	3	83	0.5	1.6	3.7	22	0.02
10 years	0.5	4	10	3	83	0.5	1.6	3.7	22	0.02
15 years	0.5	4	10	3	83	0.5	1.6	3.7	22	0.02
Adult	0.5	4	10	3	83	0.5	1.6	3.7	22	0.02

A urinary to faecal excretion ratio of 1 : 1 is assumed for technetium that has entered the transfer compartment.

Table 7.-2

Ingestion Dose Coefficients (Equivalent Dose (Sv/Bq) to Age 70 Years) for Tc-99m (T1/2 = 6.02 h)

Age at intake	3 Months	1 Year	5 Years	10 Years	15 Years	Adult
Adrenals	3.5E-11	2.0E-11	1.2E-11	7.6E-12	4.9E-12	3.9E-12
Bladder Wall	4.0E-11	3.3E-11	2.1E-11	1.5E-11	1.0E-11	7.8E-12
Bone Surfaces	4.2E-11	2.5E-11	1.3E-11	9.1E-12	6.3E-12	5.2E-12
Brain	2.3E-11	6.7E-12	3.7E-12	2.3E-12	1.4E-12	1.2E-12
Breast	2.2E-11	8.6E-12	4.6E-12	2.7E-12	1.7E-12	1.3E-12
GI-Tract						
St Wall	8.7E-10	3.2E-10	1.6E-10	9.8E-11	7.1E-11	5.6E-11
SI Wall	5.1E-11	2.2E-10	1.2E-10	7.7E-11	4.7E-11	3.8E-11
ULI Wall	1.4E-10	4.7E-10	2.5E-10	1.6E-10	9.6E-11	7.7E-11
LLI Wall	1.1E-10	3.4E-10	1.8E-10	1.1E-10	6.7E-11	5.4E-11
Kidneys	3.3E-11	2.3E-11	1.5E-11	9.9E-12	6.6E-12	5.5E-12
Liver	3.7E-11	2.7E-11	1.6E-11	9.7E-12	5.6E-12	4.4E-12
Lungs	2.9E-11	1.2E-11	6.7E-12	4.2E-12	2.7E-12	2.1E-12
Muscle	2.8E-11	1.7E-11	9.7E-12	6.4E-12	4.3E-12	3.5E-12
Ovaries	4.0E-11	7.5E-11	4.6E-11	3.1E-11	2.1E-11	1.7E-11
Pancreas	5.9E-11	3.9E-11	2.3E-11	1.7E-11	1.1E-11	9.0E-12
Red Marrow	2.5E-11	1.4E-11	9.5E-12	7.3E-12	5.2E-12	4.4E-12
Skin	2.0E-11	8.8E-12	4.7E-12	2.9E-12	1.9E-12	1.6E-12
Spleen	4.2E-11	2.6E-11	1.6E-11	1.1E-11	7.3E-12	6.3E-12
Testes	2.4E-11	1.4E-11	7.4E-12	4.5E-12	2.7E-12	2.1E-12
Thymus	2.8E-11	9.9E-12	5.1E-12	3.1E-12	2.1E-12	1.6E-12
Thyroid	1.3E-09	4.7E-10	2.5E-10	1.1E-10	7.4E-11	4.7E-11
Uterus	3.7E-11	5.5E-11	3.3E-11	2.2E-11	1.4E-11	1.1E-11
Remainder	2.8E-11	1.9E-11	1.2E-11	7.6E-12	5.1E-12	4.2E-12
Effective Dose	2.1E-10	1.3E-10	7.2E-11	4.4E-11	2.9E-11	2.2E-11

GI-Tract Gastrointestinal Tract
St Stomach
SI Small Intestine
ULI Upper Large Intestine
LLI Lower Large Intestine

Table 7.-3

Ingestion Dose Coefficients (Equivalent Dose (Sv/Bq) to Age 70 Years) for Tc-99 (T1/2 = 2.13E5 y)

Age at intake	3 Months	1 Year	5 Years	10 Years	15 Years	Adult
Adrenals	8.6E-10	2.8E-10	1.4E-10	8.2E-11	4.8E-11	3.9E-11
Bladder Wall	1.4E-09	4.9E-10	3.5E-10	2.7E-10	2.1E-10	1.6E-10
Bone Surfaces	8.6E-10	2.8E-10	1.4E-10	8.2E-11	4.8E-11	3.9E-11
Brain	8.6E-10	2.8E-10	1.4E-10	8.2E-11	4.8E-11	3.9E-11
Breast	8.6E-10	2.8E-10	1.4E-10	8.2E-11	4.8E-11	3.9E-11
GI-Tract						
St Wall	5.1E-08	1.5E-08	6.6E-09	3.8E-09	2.8E-09	2.2E-09
SI Wall	8.9E-10	1.4E-09	6.9E-10	4.1E-10	2.3E-10	1.9E-10
ULI Wall	1.1E-08	1.0E-08	5.2E-09	3.1E-09	1.7E-09	1.4E-09
LLI Wall	3.1E-08	2.9E-08	1.5E-08	8.6E-09	4.9E-09	3.9E-09
Kidneys	8.6E-10	2.8E-10	1.4E-10	8.2E-11	4.8E-11	3.9E-11
Liver	9.1E-10	3.2E-10	1.6E-10	1.0E-10	6.6E-11	5.2E-11
Lungs	8.6E-10	2.8E-10	1.4E-10	8.2E-11	4.8E-11	3.9E-11
Muscle	8.6E-10	2.8E-10	1.4E-10	8.2E-11	4.8E-11	3.9E-11
Ovaries	8.6E-10	2.8E-10	1.4E-10	8.2E-11	4.8E-11	3.9E-11
Pancreas	8.6E-10	2.8E-10	1.4E-10	8.2E-11	4.8E-11	3.9E-11
Red Marrow	8.6E-10	2.8E-10	1.4E-10	8.2E-11	4.8E-11	3.9E-11
Skin	8.6E-10	2.8E-10	1.4E-10	8.2E-11	4.8E-11	3.9E-11
Spleen	8.6E-10	2.8E-10	1.4E-10	8.2E-11	4.8E-11	3.9E-11
Testes	8.6E-10	2.8E-10	1.4E-10	8.2E-11	4.8E-11	3.9E-11
Thymus	8.6E-10	2.8E-10	1.4E-10	8.2E-11	4.8E-11	3.9E-11
Thyroid	2.5E-08	1.1E-08	5.9E-09	2.6E-09	1.6E-09	1.0E-09
Uterus	8.6E-10	2.8E-10	1.4E-10	8.2E-11	4.8E-11	3.9E-11
Remainder	8.6E-10	3.1E-10	1.5E-10	8.9E-11	5.1E-11	4.2E-11
Effective Dose	1.0E-08	4.8E-09	2.3E-09	1.3E-09	8.2E-10	6.4E-10

GI-Tract Gastrointestinal Tract
St Stomach
SI Small Intestine
ULI Upper Large Intestine
LLI Lower Large Intestine

References

Archimbaud, Y., Poncy, J.-L., Masse, R. and Cogneau, M. (1992) Etude de la toxicocinétique du technétium 95m chez le rat. *Radioprotection* **27**, 35–41.

Beasley, T. M., Palmer, H. E. and Nelp, W. B. (1966) Distribution and excretion of technetium in humans. *Health Phys.* **12**, 1425–1435.

Gerber, G. B., van Hoe, M. and Garten, C. T. (1989) Technetium absorption and turnover in monogastric and polygastric animals. *Health Phys.* **57**, 315–319.

Hays, M. T. (1972) ^{99m}Tc-pertechnetate transport in man: absorption after subcutaneous and oral administration, secretion into saliva and gastric juice. *J. Nucl. Med.* **14**, 331–335.

ICRP (1980) *Limits for Intakes of Radionuclides by Workers: Part 2,* ICRP Publication 30. Pergamon Press, Oxford.

Sullivan, M. F., Graham, T. M., Cataldo, D. A. and Schreckise, R. G. (1977) Absorption and retention of inorganic and organically incorporated technetium-95 by rats and guinea pigs. In: *Pacific Northwest Laboratory Annual Report for 1977, Part I. Biomedical Sciences,* PNL-2500, 3.95-3.101.

Sullivan, M. F., Miller, B. M. and Goebel, J. C. (1984). Gastrointestinal absorption of metals (^{51}Cr, ^{65}Zn, ^{95m}Tc, ^{109}Cd, ^{113}Sn, ^{147}Pm and ^{238}Pu) by rats and swine. *Environ. Res.* **35**, 439–453.

8. SILVER

Uptake to blood

(80) *Adults.* Furchner *et al.* (1968) reported a comparative study of the whole-body retention of silver (Ag) after intravenous and oral administration of $^{110m}AgNO_3$ in mice, rats, monkeys and dogs. Fractional absorption was less than 0.1 for silver administered as the nitrate. Based on these data ICRP (1980) adopted 0.05 as a reasonable f_1 value. Since no new data on the f_1 value seem to be available the value of 0.05 for adults is also adopted here for all compounds of silver.

(81) *Children.* Inaba *et al.* (1984) have reported data for rats on the age-dependent biokinetics of $^{110m}AgNO_3$ after oral administration. Their results indicate an enhancement of absorption of silver in suckling rats compared with adults. Following the general approach of NEA/OECD (1988) an f_1 value of 0.1 is adopted here for 3-month-old infants. For children of 1 year and older the f_1 value for the adult (0.05) is used here.

Distribution and retention

(82) *Adults.* Polachek *et al.* (1960) reported measurements of retention of ^{110m}Ag in a patient after intravenous administration as a citrate complex. Whole-body retention could be described by two components with half-times of 3.5 and 48 days. Newton and Holmes (1966) reported data on the retention of ^{110m}Ag after an accidental inhalation of a readily transportable form. A similar value for the long-term component of retention in the whole body (about 50 days) was obtained. There was a suggestion from the retention data of a small component with a longer half-time of retention.

(83) Polachek *et al.* (1960) reported *post-mortem* data on the distribution of silver in a patient at 195 days after injection; about 40% was retained in the liver, 30% in skin and the rest in other tissues. This high liver uptake is reasonably consistent with measurements reported by Newton and Holmes (1966), which indicate that following inhalation of contaminated dust from a reactor a large fraction of ^{110m}Ag entering the blood deposits in the liver, and most of it is retained with a half-time of about 50 days. Results reported by Polachek *et al.* (1960) suggest that a fraction of the deposit in other tissues is retained with a much longer half-time.

(84) Variable results have been published on the distribution of silver in experimental animals. Intraperitoneally injected silver was reported to be concentrated in the spleen and brain of rats (Furchner *et al.*, 1968), and in the spleen and liver of rats after i.v. injection (Ando *et al.*, 1989). Inhaled metallic silver was concentrated in the liver of dogs (Phalen and Morrow, 1973), and oral administration of stable silver nitrate resulted in concentration in the kidneys (Yatomi *et al.*, 1984) and liver (Kokuzawa *et al.*, 1981) of rats.

(85) It is difficult to develop a biokinetic model describing the distribution of silver from these data, although the human evidence and the animal studies indicate substantial deposition in the liver. ICRP (1980) assumed that of silver leaving the transfer compartment 0.8 is translocated to the liver and 0.2 is distributed uniformly throughout all other organs and tissues of the body including the skeleton. Of silver going to liver or any other tissue fractions, 0.1 and 0.9, are assumed to be retained with biological half-times of 3.5 and 50 days, respectively. For this report the ICRP model

has been modified to allow for the possibility of a small retention component with a long half-time. The component of retention with a 50-day half-time is therefore divided between two components with half-times of 50 days and 500 days (see Table 8.-1). It is also assumed that the initial liver deposit is 50% of activity entering blood with the remainder distributed uniformly throughout all other organs and tissues. In a single subject Newton and Holmes (1966) found no significant urinary excretion of ^{110m}Ag; however, as data are available for only one subject it is considered appropriate to allow for a small component of urinary excretion. Therefore, a urinary to faecal excretion ratio of 1:20 is assumed for silver that has entered the transfer compartment.

(86) *Children.* No information appears to be available for assessing retention and distribution in infants and children. Therefore the biokinetic data used here for adults are also adopted for infants and children.

Dose coefficients

(87) Dose coefficients derived from the biokinetic data summarized in Table 8.-1 are given in Tables 8.-2 and 8.-3.

Table 8.-1. Biokinetic data for silver

		Distribution (%)		Biological half-time (days)		
Age	f_1	Liver	Other tissues	Comp. A (10%)	Comp. B (80%)	Comp. C (10%)
3 months	0.1	50	50	3.5	50	500
1 year	0.05	50	50	3.5	50	500
5 years	0.05	50	50	3.5	50	500
10 years	0.05	50	50	3.5	50	500
15 years	0.05	50	50	3.5	50	500
Adult	0.05	50	50	3.5	50	500

A urinary to faecal excretion ratio of 1:20 is assumed for silver that has entered the transfer compartment.

Table 8.-2

Ingestion Dose Coefficients (Equivalent Dose (Sv/Bq) to Age 70 Years) for Ag-108m (T1/2 = 127 y)

Age at intake	3 Months	1 Year	5 Years	10 Years	15 Years	Adult
Adrenals	1.6E-08	6.9E-09	4.6E-09	3.3E-09	2.3E-09	2.0E-09
Bladder Wall	9.9E-09	5.3E-09	3.2E-09	2.1E-09	1.3E-09	1.1E-09
Bone Surfaces	9.0E-09	3.9E-09	2.2E-09	1.5E-09	1.0E-09	8.9E-10
Brain	4.6E-09	1.9E-09	1.1E-09	6.9E-10	4.5E-10	3.9E-10
Breast	7.1E-09	3.0E-09	1.7E-09	1.1E-09	7.2E-10	6.1E-10
GI-Tract						
St Wall	1.8E-08	8.4E-09	4.6E-09	2.8E-09	1.8E-09	1.5E-09
SI Wall	2.3E-08	1.3E-08	7.8E-09	5.1E-09	3.2E-09	2.7E-09
ULI Wall	4.2E-08	2.6E-08	1.4E-08	9.3E-09	5.7E-09	4.7E-09
LLI Wall	7.2E-08	4.8E-08	2.7E-08	1.7E-08	1.0E-08	8.6E-09
Kidneys	1.4E-08	6.2E-09	4.1E-09	2.8E-09	1.9E-09	1.6E-09
Liver	7.6E-08	3.2E-08	2.0E-08	1.4E-08	1.0E-08	8.9E-09
Lungs	1.1E-08	4.6E-09	2.8E-09	1.9E-09	1.3E-09	1.1E-09
Muscle	8.5E-09	3.9E-09	2.3E-09	1.5E-09	1.0E-09	8.9E-10
Ovaries	2.0E-08	1.3E-08	7.3E-09	4.9E-09	3.3E-09	2.5E-09
Pancreas	2.0E-08	8.7E-09	5.3E-09	3.4E-09	2.3E-09	1.9E-09
Red Marrow	7.1E-09	3.3E-09	2.3E-09	1.7E-09	1.3E-09	1.1E-09
Skin	5.5E-09	2.3E-09	1.4E-09	8.7E-10	5.9E-10	5.2E-10
Spleen	1.0E-08	4.5E-09	2.7E-09	1.7E-09	1.1E-09	9.1E-10
Testes	6.6E-09	3.3E-09	1.8E-09	1.2E-09	7.2E-10	6.3E-10
Thymus	6.9E-09	2.9E-09	1.8E-09	1.1E-09	8.1E-10	6.7E-10
Thyroid	6.4E-09	2.6E-09	1.5E-09	9.3E-10	6.1E-10	5.2E-10
Uterus	1.4E-08	7.7E-09	4.5E-09	2.9E-09	1.8E-09	1.5E-09
Remainder	8.0E-09	3.7E-09	2.3E-09	1.5E-09	1.1E-09	9.1E-10
Effective Dose	2.1E-08	1.1E-08	6.6E-09	4.3E-09	2.9E-09	2.4E-09

GI-Tract Gastrointestinal Tract
St Stomach
SI Small Intestine
ULI Upper Large Intestine
LLI Lower Large Intestine

Table 8.-3

Ingestion Dose Coefficients (Equivalent Dose (Sv/Bq) to Age 70 Years) for Ag-110m (T1/2 = 249.9 d)

Age at intake	3 Months	1 Year	5 Years	10 Years	15 Years	Adult
Adrenals	1.6E-08	7.2E-09	4.5E-09	3.3E-09	2.2E-09	1.9E-09
Bladder Wall	1.2E-08	7.0E-09	4.1E-09	2.7E-09	1.6E-09	1.4E-09
Bone Surfaces	9.3E-09	4.2E-09	2.3E-09	1.5E-09	1.0E-09	8.6E-10
Brain	4.6E-09	1.8E-09	1.0E-09	6.5E-10	4.0E-10	3.4E-10
Breast	7.5E-09	3.1E-09	1.8E-09	1.1E-09	7.1E-10	5.7E-10
GI-Tract						
St Wall	2.0E-08	9.9E-09	5.4E-09	3.3E-09	2.1E-09	1.7E-09
SI Wall	2.8E-08	1.7E-08	1.0E-08	6.8E-09	4.3E-09	3.5E-09
ULI Wall	5.3E-08	3.3E-08	1.9E-08	1.2E-08	7.6E-09	6.3E-09
LLI Wall	9.3E-08	6.3E-08	3.5E-08	2.2E-08	1.4E-08	1.2E-08
Kidneys	1.5E-08	6.7E-09	4.3E-09	2.9E-09	1.9E-09	1.6E-09
Liver	7.1E-08	3.0E-08	1.8E-08	1.3E-08	9.0E-09	7.6E-09
Lungs	1.1E-08	4.7E-09	2.7E-09	1.8E-09	1.3E-09	9.9E-10
Muscle	9.4E-09	4.5E-09	2.6E-09	1.7E-09	1.1E-09	9.5E-10
Ovaries	2.6E-08	1.7E-08	1.0E-08	6.7E-09	4.6E-09	3.5E-09
Pancreas	2.1E-08	9.2E-09	5.5E-09	3.6E-09	2.3E-09	1.8E-09
Red Marrow	7.6E-09	3.7E-09	2.6E-09	1.9E-09	1.4E-09	1.2E-09
Skin	6.1E-09	2.7E-09	1.5E-09	9.2E-10	6.3E-10	5.4E-10
Spleen	1.1E-08	5.1E-09	2.9E-09	1.9E-09	1.2E-09	9.4E-10
Testes	7.7E-09	3.9E-09	2.2E-09	1.4E-09	8.4E-10	6.9E-10
Thymus	7.1E-09	2.9E-09	1.7E-09	1.1E-09	7.7E-10	6.3E-10
Thyroid	6.5E-09	2.6E-09	1.5E-09	9.0E-10	5.6E-10	4.6E-10
Uterus	1.7E-08	1.0E-08	5.8E-09	3.8E-09	2.4E-09	1.9E-09
Remainder	8.8E-09	4.2E-09	2.5E-09	1.7E-09	1.2E-09	9.8E-10
Effective Dose	2.4E-08	1.4E-08	7.9E-09	5.2E-09	3.4E-09	2.8E-09

GI-Tract	Gastrointestinal Tract
St	Stomach
SI	Small Intestine
ULI	Upper Large Intestine
LLI	Lower Large Intestine

References

Ando, A., Ando, I., Hiraki, T. and Hisada, K. (1989) Relation between the location of elements in the Periodic Table and various organ-uptake rates. *Nucl. Med. Biol.* **16**, 57–80.

Furchner, J. E., Richmond, C. R. and Drake, G. A. (1968) Comparative metabolism of radionuclides in mammals—IV. Retention of silver-110m in the mouse, rat, monkey and dog. *Health Phys.* **15**, 505–514.

ICRP (1980) *Limits for Intakes of Radionuclides by Workers: Part 2,* ICRP Publication 30. Pergamon Press, Oxford.

Inaba, J., Nishimura, Y. and Ichikawa, R. (1984) Effect of age on the metabolism of some important radionuclides in the rat. In: *Radiation Risk Protection. Proc. 6th Int. Congress of IRPA* (A. Kaul, R. Neider, J. Pensko, F.-E. Stieve and H. Brunner, Eds), Vol. 1, pp. 481–484. Fachverband für Strahlenschutz e.V., Berlin.

Kokuzawa, S., Nose, T., Sugiyama, K., Tanaka, T. and Ohshiro, H. (1981) Experimental study on the interaction of Ag and Cu in the rat. *Jap. J. Hyg.* **36**(1), 301.

NEA/OECD (1988) Committee on Radiation Protection and Public Health. *Report of an Expert Group on Gut Transfer Factors.* NEA/OECD Report, Paris.

Newton, D. and Holmes, A. (1966) A case of accidental inhalation of zinc-65 and silver-110m. *Radiat. Res.* **29**, 403–412.

Phalen, R. F. and Morrow, P. E. (1973) Experimental inhalation of metallic silver. *Health Phys.* **24**, 509–518.

Polachek, A. A., Cope, C. B., Williard, R. F. and Enns, T. (1960) Metabolism of radioactive silver in a patient with carcinoid. *J. Lab. Clin. Med.* **56**, 499–505.

Yatomi, H., Sugiyama, K., Iizuka, S., Kurosawa, Y., Tanaka, T., Funakawa, K. and Nose, T. (1984) Distribution of Ag, Cu and Zn after chronical oral administration of silver in rats. *Jap. J. Hyg.* **39**(1), 504.

9. TELLURIUM

Uptake to blood

(88) *Adults.* There appears to be no information available for humans on the absorption of tellurium (Te). Furthermore, the metabolism of tellurium has not been studied extensively in other mammals. Venugopal and Luckey (1978) reviewed experimental data from several animal species which indicate that about 25% of ingested water-soluble tellurites (TeO_2) and tellurates (TeO_3) are absorbed. Hollins (1969) reported fractional gastrointestinal absorption of ^{127m}Te in the range 0.1–0.16 for orally administered tellurate in rats. DeMeio and Henriques (1947) gave f_1 values of 0.2–0.5 for sodium tellurite in dogs and Wright and Bell (1966) reported f_1 values for tellurium of 0.24 in sheep and 0.25 in swine following administration of sodium tellurite. Mullen and Stanley (1974), in a study of the biokinetics of tellurium in dairy cows, reported that approximately one-quarter of ^{129m}Te given orally as tellurate was absorbed from the gastrointestinal tract.

(89) In *ICRP Publication 30* (ICRP, 1979) an f_1 value of 0.2 was adopted for workers. The data summarized above suggest that 0.3 is a more appropriate f_1 value for tellurium ingested in diet by adult members of the public, and this value is adopted here.

(90) *Children.* No information appears to be available from experimental animals or studies in humans for assessing any possible changes with age in the f_1 value for tellurium. Following the general approach of NEA/OECD (1988) an f_1 value for tellurium of 0.6 is adopted here for the 3-month-old infant. For children of 1 year and older the f_1 value for adults (0.3) is used.

Distribution and retention

(91) *Adults.* There are limited published data on the distribution and retention of tellurium in humans and animals from which to develop a biokinetic model. Hollins (1969) studied the retention of ^{127m}Te in rats after intraperitoneal administration of tellurate and obtained two components of retention with half-times of about 0.8 days (49%) and 13 days (51%). Highest concentrations were found in the kidneys up to 20 days after administration, with levels about 10 times the average concentration in the body. At 200 days the highest concentration of ^{127m}Te was found in the skeleton, being 20 times the average concentration in the body; accumulation in the thyroid was not measured. Casey *et al.* (1963) studied the biokinetics of orally administered ^{132}Te in lactating sheep up to 96 h after administration. The concentration in the thyroid and kidneys was about 15–20 times the average in other tissues at 3–4 days after administration; retention in bone was not determined.

(92) The following model for tellurium is adopted for the calculation of dose coefficients for adults. It is based on that given in *ICRP Publication 30* (ICRP, 1979), but allows for the higher concentrations initially deposited in the kidneys and thyroid. It is assumed that 50% of tellurium entering the transfer compartment goes directly to excretion with a half-time of 0.8 days. Twenty-five per cent is assumed to be translocated to the skeleton (ICRP, 1979). According to the general approach of paragraph (7) in the Introduction, a biological half-time in the skeleton of 10 000 days is assumed. This value does not have a significant effect on the calculation of dose coefficients as the principle

isotopes of tellurium formed in fission have short half-lives. A further 0.2% is assumed to go to the thyroid, 2.3% to the kidneys and 22.5% to the remainder tissues where it is retained with a biological half-time of 20 days. A urinary to faecal excretion ratio of 4:1 is assumed for tellurium that has entered the transfer compartment (Hollins, 1969).

(93) *Children.* There appear to be no data on age-related changes in the distribution and retention of tellurium in either animals or man. The retention parameters adopted here for adults are also taken to apply to infants and children.

Classification of isotopes for bone dosimetry

(94) No data have been found concerning the location of isotopes of tellurium in the skeleton. Because many of these isotopes have short radioactive half-lives, it is assumed for purposes of dosimetry that all isotopes of tellurium are retained on bone surfaces (ICRP, 1979).

Behaviour of daughters

(95) It is assumed that all radioisotopes of iodine produced by the decay of systemic tellurium are translocated instantaneously to the transfer compartment in inorganic form and have the same biokinetic behaviour as systemic iodine. Radioiodines produced within the thyroid are retained as organic iodine. Radioiodines produced within the stomach and small intestine are absorbed according to normal gastrointestinal (g.i) tract biokinetics, but decay products formed in the lower segments of the g.i tract are excreted.

Dose coefficients

Dose coefficients derived from the biokinetic data summarized in Table 9.-1 are given in Tables 9.-2 to 9.-5.

Table 9.-1. Biokinetic data for tellurium

Age	f_1	Distribution (%)					Biological half-time (days)		
		Skeleton	Thyroid	Kidney	Other tissues	Prompt excretion	Skeleton	Tissues other than skeleton	Transfer comp.
3 months	0.6	25	0.2	2.3	22.5	50	10 000	20	0.8
1 year	0.3	25	0.2	2.3	22.5	50	10 000	20	0.8
5 years	0.3	25	0.2	2.3	22.5	50	10 000	20	0.8
10 years	0.3	25	0.2	2.3	22.5	50	10 000	20	0.8
15 years	0.3	25	0.2	2.3	22.5	50	10 000	20	0.8
Adult	0.3	25	0.2	2.3	22.5	50	10 000	20	0.8

A urinary to faecal excretion ratio of 4:1 is assumed for tellurium that has entered the transfer compartment.

Table 9.-2

Ingestion Dose Coefficients (Equivalent Dose (Sv/Bq) to Age 70 Years) for Te-127m (T1/2 = 109 d)

Age at intake	3 Months	1 Year	5 Years	10 Years	15 Years	Adult
Adrenals	3.0E-09	1.0E-09	5.1E-10	3.1E-10	1.8E-10	1.5E-10
Bladder Wall	3.7E-09	1.3E-09	8.7E-10	6.3E-10	4.7E-10	3.6E-10
Bone Surfaces	3.3E-07	1.4E-07	1.0E-07	5.5E-08	3.2E-08	3.2E-08
Brain	2.9E-09	9.9E-10	4.9E-10	2.9E-10	1.8E-10	1.4E-10
Breast	2.8E-09	9.3E-10	4.6E-10	2.8E-10	1.6E-10	1.3E-10
GI-Tract						
St Wall	4.3E-09	1.8E-09	8.6E-10	5.0E-10	3.2E-10	2.5E-10
SI Wall	4.6E-09	3.1E-09	1.6E-09	9.4E-10	5.3E-10	4.3E-10
ULI Wall	2.5E-08	2.3E-08	1.1E-08	6.8E-09	3.8E-09	3.0E-09
LLI Wall	8.0E-08	7.7E-08	3.9E-08	2.3E-08	1.3E-08	1.0E-08
Kidneys	3.4E-08	1.1E-08	6.1E-09	4.1E-09	2.9E-09	2.3E-09
Liver	2.8E-09	9.5E-10	4.7E-10	2.8E-10	1.7E-10	1.4E-10
Lungs	2.9E-09	9.7E-10	4.8E-10	2.9E-10	1.7E-10	1.4E-10
Muscle	2.9E-09	9.7E-10	4.8E-10	2.9E-10	1.7E-10	1.4E-10
Ovaries	3.1E-09	1.2E-09	5.9E-10	3.6E-10	2.1E-10	1.7E-10
Pancreas	2.9E-09	9.6E-10	4.8E-10	2.9E-10	1.7E-10	1.4E-10
Red Marrow	2.1E-07	7.4E-08	3.7E-08	1.9E-08	1.1E-08	8.2E-09
Skin	2.8E-09	9.3E-10	4.6E-10	2.8E-10	1.6E-10	1.3E-10
Spleen	2.9E-09	9.6E-10	4.8E-10	2.9E-10	1.7E-10	1.4E-10
Testes	2.8E-09	9.3E-10	4.6E-10	2.7E-10	1.6E-10	1.3E-10
Thymus	2.8E-09	9.4E-10	4.6E-10	2.8E-10	1.7E-10	1.4E-10
Thyroid	7.6E-08	3.4E-08	1.7E-08	7.7E-09	4.9E-09	3.1E-09
Uterus	2.9E-09	1.0E-09	5.0E-10	3.0E-10	1.7E-10	1.4E-10
Remainder	3.5E-09	1.2E-09	6.1E-10	3.5E-10	2.1E-10	1.7E-10
Effective Dose	4.1E-08	1.8E-08	9.5E-09	5.2E-09	3.0E-09	2.3E-09

GI-Tract Gastrointestinal Tract
St Stomach
SI Small Intestine
ULI Upper Large Intestine
LLI Lower Large Intestine

Table 9.-3

Ingestion Dose Coefficients (Equivalent Dose (Sv/Bq) to Age 70 Years) for Te-129m (T1/2 = 33.6 d)

Age at intake	3 Months	1 Year	5 Years	10 Years	15 Years	Adult
Adrenals	4.9E-09	1.7E-09	8.3E-10	5.1E-10	3.1E-10	2.5E-10
Bladder Wall	6.6E-09	2.4E-09	1.6E-09	1.2E-09	9.2E-10	7.0E-10
Bone Surfaces	1.4E-07	5.7E-08	4.0E-08	2.2E-08	1.3E-08	1.2E-08
Brain	4.7E-09	1.6E-09	7.8E-10	4.7E-10	2.8E-10	2.3E-10
Breast	4.6E-09	1.5E-09	7.5E-10	4.5E-10	2.7E-10	2.2E-10
GI-Tract						
St Wall	1.1E-08	4.8E-09	2.3E-09	1.4E-09	8.8E-10	7.0E-10
SI Wall	1.2E-08	1.0E-08	5.0E-09	3.0E-09	1.7E-09	1.4E-09
ULI Wall	6.4E-08	6.1E-08	3.0E-08	1.8E-08	1.0E-08	8.1E-09
LLI Wall	1.7E-07	1.7E-07	8.4E-08	5.0E-08	2.8E-08	2.3E-08
Kidneys	5.2E-08	1.7E-08	9.2E-09	6.2E-09	4.3E-09	3.5E-09
Liver	4.7E-09	1.6E-09	7.9E-10	4.8E-10	2.8E-10	2.3E-10
Lungs	4.7E-09	1.6E-09	7.8E-10	4.7E-10	2.8E-10	2.3E-10
Muscle	4.7E-09	1.6E-09	7.9E-10	4.8E-10	2.9E-10	2.3E-10
Ovaries	5.1E-09	2.0E-09	1.0E-09	6.4E-10	3.9E-10	3.1E-10
Pancreas	4.8E-09	1.6E-09	8.0E-10	4.9E-10	2.9E-10	2.4E-10
Red Marrow	1.6E-07	5.1E-08	2.4E-08	1.3E-08	7.5E-09	5.3E-09
Skin	4.6E-09	1.5E-09	7.6E-10	4.5E-10	2.7E-10	2.2E-10
Spleen	4.8E-09	1.6E-09	8.0E-10	4.8E-10	2.9E-10	2.3E-10
Testes	4.6E-09	1.6E-09	7.7E-10	4.6E-10	2.8E-10	2.2E-10
Thymus	4.6E-09	1.5E-09	7.6E-10	4.6E-10	2.8E-10	2.2E-10
Thyroid	1.1E-07	5.1E-08	2.6E-08	1.1E-08	7.4E-09	4.6E-09
Uterus	4.8E-09	1.7E-09	8.7E-10	5.3E-10	3.2E-10	2.6E-10
Remainder	5.8E-09	2.0E-09	1.0E-09	6.1E-10	3.6E-10	2.9E-10
Effective Dose	4.4E-08	2.4E-08	1.2E-08	6.6E-09	3.9E-09	3.0E-09

GI-Tract	Gastrointestinal Tract
St	Stomach
SI	Small Intestine
ULI	Upper Large Intestine
LLI	Lower Large Intestine

Table 9.-4

Ingestion Dose Coefficients (Equivalent Dose (Sv/Bq) to Age 70 Years) for Te-131m (T1/2 = 30 h)

Age at intake	3 Months	1 Year	5 Years	10 Years	15 Years	Adult
Adrenals	1.4E-09	6.9E-10	3.8E-10	2.4E-10	1.6E-10	1.2E-10
Bladder Wall	2.5E-09	1.6E-09	1.1E-09	7.9E-10	5.4E-10	4.4E-10
Bone Surfaces	5.8E-09	2.4E-09	1.6E-09	8.9E-10	5.3E-10	5.0E-10
Brain	9.2E-10	3.1E-10	1.7E-10	1.1E-10	6.7E-11	5.7E-11
Breast	8.9E-10	3.6E-10	1.9E-10	1.2E-10	7.0E-11	5.7E-11
GI-Tract						
St Wall	6.7E-09	4.0E-09	2.1E-09	1.3E-09	8.6E-10	6.8E-10
SI Wall	7.8E-09	8.2E-09	4.5E-09	2.8E-09	1.7E-09	1.4E-09
ULI Wall	2.9E-08	3.0E-08	1.6E-08	9.5E-09	5.5E-09	4.4E-09
LLI Wall	5.3E-08	5.4E-08	2.8E-08	1.7E-08	9.8E-09	7.9E-09
Kidneys	3.2E-09	1.4E-09	8.1E-10	5.4E-10	3.6E-10	3.0E-10
Liver	1.4E-09	8.5E-10	4.7E-10	2.9E-10	1.7E-10	1.3E-10
Lungs	1.1E-09	4.6E-10	2.4E-10	1.5E-10	9.5E-11	7.4E-11
Muscle	1.3E-09	7.2E-10	4.0E-10	2.6E-10	1.7E-10	1.4E-10
Ovaries	3.6E-09	3.5E-09	2.0E-09	1.4E-09	9.6E-10	7.3E-10
Pancreas	1.6E-09	9.5E-10	5.4E-10	3.5E-10	2.2E-10	1.8E-10
Red Marrow	3.5E-09	1.3E-09	7.3E-10	4.8E-10	3.3E-10	2.7E-10
Skin	9.3E-10	4.2E-10	2.3E-10	1.4E-10	9.0E-11	7.5E-11
Spleen	1.4E-09	7.7E-10	4.4E-10	2.8E-10	1.7E-10	1.4E-10
Testes	1.2E-09	7.1E-10	3.9E-10	2.4E-10	1.4E-10	1.1E-10
Thymus	1.1E-09	4.3E-10	2.3E-10	1.4E-10	8.6E-11	6.8E-11
Thyroid	2.6E-07	1.5E-07	8.8E-08	4.5E-08	2.9E-08	1.9E-08
Uterus	2.4E-09	2.1E-09	1.2E-09	7.8E-10	4.8E-10	3.8E-10
Remainder	1.4E-09	8.1E-10	4.7E-10	3.0E-10	2.0E-10	1.6E-10
Effective Dose	2.0E-08	1.4E-08	7.8E-09	4.3E-09	2.7E-09	2.0E-09

GI-Tract Gastrointestinal Tract
- St Stomach
- SI Small Intestine
- ULI Upper Large Intestine
- LLI Lower Large Intestine

Table 9.-5

Ingestion Dose Coefficients (Equivalent Dose (Sv/Bq) to Age 70 Years)
for Te-132 (T1/2 = 78.2 h)

Age at intake	3 Months	1 Year	5 Years	10 Years	15 Years	Adult
Adrenals	3.8E-09	1.7E-09	9.5E-10	5.9E-10	3.8E-10	3.1E-10
Bladder Wall	1.6E-08	8.4E-09	6.3E-09	5.1E-09	4.1E-09	3.2E-09
Bone Surfaces	1.3E-08	5.6E-09	3.6E-09	2.1E-09	1.2E-09	1.2E-09
Brain	3.0E-09	1.1E-09	6.0E-10	3.7E-10	2.3E-10	2.0E-10
Breast	2.8E-09	1.1E-09	5.8E-10	3.6E-10	2.2E-10	1.8E-10
GI-Tract						
St Wall	7.8E-09	4.4E-09	2.4E-09	1.4E-09	9.6E-10	7.7E-10
SI Wall	9.0E-09	7.9E-09	4.6E-09	3.0E-09	1.9E-09	1.5E-09
ULI Wall	5.0E-08	5.1E-08	2.6E-08	1.6E-08	9.3E-09	7.5E-09
LLI Wall	1.3E-07	1.3E-07	6.8E-08	4.1E-08	2.4E-08	1.9E-08
Kidneys	6.0E-09	2.6E-09	1.5E-09	9.8E-10	6.5E-10	5.3E-10
Liver	3.9E-09	2.0E-09	1.1E-09	6.7E-10	4.0E-10	3.3E-10
Lungs	3.3E-09	1.4E-09	7.1E-10	4.5E-10	2.8E-10	2.3E-10
Muscle	3.8E-09	1.9E-09	1.1E-09	6.9E-10	4.5E-10	3.7E-10
Ovaries	8.8E-09	7.9E-09	4.6E-09	3.1E-09	2.2E-09	1.6E-09
Pancreas	4.1E-09	2.1E-09	1.2E-09	7.4E-10	4.6E-10	3.7E-10
Red Marrow	6.4E-09	2.6E-09	1.5E-09	1.0E-09	6.9E-10	5.6E-10
Skin	2.8E-09	1.2E-09	6.6E-10	4.1E-10	2.6E-10	2.2E-10
Spleen	3.8E-09	1.9E-09	1.0E-09	6.5E-10	4.0E-10	3.3E-10
Testes	3.9E-09	2.1E-09	1.2E-09	7.5E-10	4.4E-10	3.6E-10
Thymus	3.8E-09	1.5E-09	7.6E-10	4.6E-10	2.9E-10	2.3E-10
Thyroid	6.2E-07	3.0E-07	1.6E-07	7.1E-08	4.6E-08	2.9E-08
Uterus	6.2E-09	4.5E-09	2.7E-09	1.8E-09	1.1E-09	8.8E-10
Remainder	3.7E-09	1.9E-09	1.1E-09	7.1E-10	4.7E-10	3.9E-10
Effective Dose	4.6E-08	2.9E-08	1.5E-08	8.1E-09	5.1E-09	3.7E-09

GI-Tract Gastrointestinal Tract
St Stomach
SI Small Intestine
ULI Upper Large Intestine
LLI Lower Large Intestine

References

Casey, H. W., Case, A. C., McClellan, R. O. and Bustad, L. K. (1963) Metabolism of Te-132–I-132 in lactating sheep. *Health Phys.* **9**, 1223–1225.

DeMeio, R. H. and Henriques, F. C. Jr. (1947) Excretion and distribution in tissues studied with a radioactive isotope. *J. Biol. Chem.* **169**, 609–617.

Hollins, J. G. (1969) The metabolism of tellurium in rats. *Health Phys.* **17**, 497–505.

ICRP (1979) *Limits for Intake of Radionuclides by Workers: Part 1,* ICRP Publication 30. Pergamon Press, Oxford.

ICRP (1989) *Age-dependent Doses to Members of the Public from Intake of Radionuclides: Part 1,* ICRP Publication 56. Pergamon Press, Oxford.

Mullen, A. A. and Stanley, R. E. (1974) Absorption, distribution and milk secretion of radionuclides by the dairy cow, III, radiotellurium. *Health Phys.* **27**, 279–284.

NEA/OECD (1988) Committee on Radiation Protection and Public Health. *Report of an Expert Group on Gut Transfer Factors.* NEA/OECD Report, Paris.

Venugopal, B. and Luckey, T. D. (1978) *Metal Toxicity in Mammals.* Vol. 2. Plenum Press, New York.

Wright, P. L. and Bell, M. C. (1966) Comparative metabolism of selenium and tellurium in sheep and swine. *Am. J. Physiol.* **20**, 6–10.

10. BARIUM

Uptake to blood

(96) *Adults.* The fractional absorption of barium (Ba) has been reviewed by Leggett (1992b). Its absorption depends on its chemical form. Barium sulphate is very poorly absorbed from the gastrointestinal tract of adults (Figueroa *et al.*, 1968; Boender and Verloop, 1969), but acid-soluble barium salts (e.g. acetate, carbonate, chloride, hydroxide, nitrate and sulphide) are readily dissolved in gastric acid and absorbed. In five female cancer patients with normal gastrointestinal function, 3–16% (mean $8 \pm 6\%$) of $^{140}BaCl_2$ administered in orange juice was absorbed (Bligh, 1960). In several subjects ingesting ^{133}Ba as simulated nuclear weapon fallout, estimates of absorption varied from immeasurably small amounts to 15%, and averaged less than 6% (LeRoy *et al.*, 1966). However, these estimates of absorption were extrapolated from retention measurements of ^{133}Ba made several days after intake, and would be higher if endogenous secretion during the first few days after ingestion of barium had been considered (Harrison *et al.*, 1967; Korsunskii *et al.*, 1981). A recalculation of the results of a balance study of three humans (Tipton *et al.*, 1969; Schroeder *et al.*, 1972) gave estimated fractional absorption values of 20%, 20% and 60%, respectively, assuming a urinary to faecal excretion ratio from systemic activity of 1:9 (Harrison *et al.*, 1967; Newton *et al.*, 1991), but estimates of absorption based on balance studies are highly uncertain. Results of studies on the comparative metabolism of ingested barium and radium in rats, dogs, sheep, pigs and cows indicate that fractional absorption is similar for these two elements (Garner *et al.*, 1960; Taylor *et al.*, 1962; Sansom and Garner, 1966; Della Rosa *et al.*, 1967; Hardy *et al.*, 1969).

(97) In *ICRP Publication 30* (ICRP, 1979), fractional absorption from the gastrointestinal tract is assumed to be 0.1 for all compounds of barium. However, the available human data are variable and indicate that greater absorption may occur (Leggett, 1992b). In view of the chemical similarities of barium and radium, the similar results obtained for their absorption in experimental animals and the adoption of an f_1 of 0.2 for radium, an f_1 value for barium of 0.2 is also adopted here.

(98) *Children.* There appear to be no direct measurements of barium absorption in children. However, data from animals indicate that the gastrointestinal absorption of barium is greater in immature than in mature animals, as is the case for other alkaline earth elements. In rats, there was an inverse relationship of barium absorption with age after administration as the chloride (Taylor *et al.*, 1962). In suckling rats 14–18 days of age, the absorption of both barium and radium was about 80% and in young adult rats (6–8 weeks) about 10%. These results suggest that the high fractional absorption in suckling infants decreases rapidly with increasing age. Della Rosa *et al.* (1967) reported that following oral administration of ^{133}Ba to beagle dogs of 43, 150 and 250 days of age retention at 30 days after administration was 2.3%, 2.0% and 0.8%, respectively, and 0.4–0.6% in adult dogs. Cuddihy and Griffith (1972) estimated from these data that gastrointestinal absorption may have been 0.7–1.5% in the adult dogs and as high as 7% in the younger animals.

(99) In the present document, a fractional absorption of barium of 0.6 is adopted for the infant. For ages 1–15 years, a value of 0.3 is used. The choice of this value is

based on the consideration that there may be elevated absorption of barium throughout the period of growth.

Distribution and retention

(100) *Adults.* The distribution, retention and excretion of injected or ingested radiobarium have been studied in several human subjects, primarily adults (Bauer *et al.*, 1957; LeRoy *et al.*, 1966; Harrison *et al.*, 1967; Rundo, 1967; Newton *et al.*, 1977; Erre *et al.*, 1980; Korsunskii *et al.*, 1981; Harrison, 1981; Newton *et al.*, 1991). Injected radiobarium is rapidly cleared from the circulation. It can be inferred from the data and analyses of Bauer *et al.* (1957), Harrison *et al.* (1967), Newton *et al.* (1967), Erre *et al.* (1980), Korsunskii *et al.* (1981) and Newton *et al.* (1991) that the skeletal retention of barium in an adult human decreases from one-quarter or more of injected activity in the first day or two after injection to approximately 10% after 1 month and 5% after 1 year. Some barium is also retained in soft tissues (Hardy *et al.*, 1969; Schroeder *et al.*, 1972; van Middlesworth and Robinson, 1975).

(101) The age-specific model (Leggett, 1992) adopted here for barium was designed to yield predictions of retention in adults that are slightly higher than predictions of the model developed by the Task Group on Alkaline Earth Metabolism in Adult Man (ICRP, 1973), in view of recent data on injected barium in adult volunteers (Newton *et al.*, 1991). The present model also depicts more explicitly the anatomical compartments and physiological processes needed to model age-specific biokinetics of barium. This model is described in Appendix A.

(102) *Children.* Age dependence in the biokinetics of barium has been investigated in laboratory animals (Della Rosa *et al.*, 1967; Ellsasser *et al.*, 1969; Hardy *et al.*, 1969; Cuddihy and Griffith, 1972; Stather, 1974; Domanski *et al.*, 1980) and in one injection study involving human infants, children and adults (Bauer *et al.*, 1957). Turnover time in the rapidly exchanging pools including plasma and extracellular fluids appears to be substantially greater in immature than mature animals (Ellsasser *et al.*, 1969), and the rate of transfer from blood to bone is higher in immature animals (Bauer *et al.*, 1957; Ellsasser *et al.*, 1969). The choice of age-specific biokinetic parameters for the model adopted here for barium is described in Appendix A.

Dose coefficients

(103) Dose coefficients derived from the biokinetic data summarized in Table A-2 are given in Tables 10.-1 and 10.-2.

Table 10.-1

Ingestion Dose Coefficients (Equivalent Dose (Sv/Bq) to Age 70 Years) for Ba-133 (T1/2 = 10.74 y)*

Age at intake	3 Months	1 Year	5 Years	10 Years	15 Years	Adult
Adrenals	1.8E-08	3.9E-09	2.6E-09	4.2E-09	7.8E-09	1.2E-09
Bladder Wall	8.4E-09	2.5E-09	1.7E-09	2.1E-09	2.9E-09	6.2E-10
Bone Surfaces	1.5E-07	2.7E-08	1.9E-08	2.9E-08	5.3E-08	6.6E-09
Brain	1.2E-08	2.6E-09	2.1E-09	3.5E-09	7.0E-09	8.4E-10
Breast	6.9E-09	1.5E-09	9.9E-10	1.4E-09	2.3E-09	3.6E-10
GI-Tract						
St Wall	9.0E-09	2.7E-09	1.7E-09	1.9E-09	2.9E-09	6.5E-10
SI Wall	1.3E-08	5.1E-09	3.2E-09	3.5E-09	5.1E-09	1.3E-09
ULI Wall	2.2E-08	1.2E-08	6.8E-09	5.5E-09	5.7E-09	2.3E-09
LLI Wall	4.7E-08	2.8E-08	1.5E-08	1.1E-08	1.0E-08	4.9E-09
Kidneys	1.3E-08	2.9E-09	1.9E-09	2.8E-09	4.9E-09	8.2E-10
Liver	9.0E-09	2.2E-09	1.4E-09	1.9E-09	3.3E-09	5.5E-10
Lungs	1.2E-08	2.5E-09	1.7E-09	2.5E-09	4.5E-09	6.6E-10
Muscle	1.2E-08	2.6E-09	1.8E-09	2.6E-09	4.7E-09	6.8E-10
Ovaries	1.4E-08	5.1E-09	3.3E-09	3.9E-09	5.8E-09	1.4E-09
Pancreas	1.1E-08	2.6E-09	1.7E-09	2.5E-09	4.5E-09	7.7E-10
Red Marrow	6.6E-08	1.2E-08	7.8E-09	1.2E-08	2.2E-08	3.7E-09
Skin	9.0E-09	1.9E-09	1.2E-09	1.8E-09	3.1E-09	4.3E-10
Spleen	1.0E-08	2.4E-09	1.5E-09	2.1E-09	3.4E-09	5.6E-10
Testes	6.9E-09	1.8E-09	1.1E-09	1.5E-09	2.4E-09	3.8E-10
Thymus	8.1E-09	1.8E-09	1.2E-09	1.8E-09	3.3E-09	5.1E-10
Thyroid	7.6E-09	1.8E-09	1.4E-09	2.2E-09	4.3E-09	6.0E-10
Uterus	9.9E-09	3.2E-09	2.1E-09	2.5E-09	4.0E-09	8.4E-10
Remainder	1.2E-08	2.7E-09	1.8E-09	2.7E-09	4.8E-09	7.1E-10
Effective Dose	2.1E-08	6.2E-09	3.9E-09	4.7E-09	7.2E-09	1.5E-09

GI-Tract Gastrointestinal Tract
St Stomach
SI Small Intestine
ULI Upper Large Intestine
LLI Lower Large Intestine

* In the biokinetic model for Ba parameter values for the adult apply to ages > 25 y. For radioisotopes of this element the dose coefficients for the adult are based on the 50-y integrated doses following an acute intake at age 25 y.

Table 10.-2

Ingestion Dose Coefficients (Equivalent Dose (Sv/Bq) to Age 70 Years) for Ba-140 (T1/2 = 12.74 d)*

Age at intake	3 Months	1 Year	5 Years	10 Years	15 Years	Adult
Adrenals	8.6E-09	2.0E-09	9.7E-10	9.0E-10	9.3E-10	2.1E-10
Bladder Wall	5.4E-09	2.5E-09	1.4E-09	1.1E-09	7.1E-10	4.5E-10
Bone Surfaces	1.0E-07	2.0E-08	1.2E-08	1.1E-08	9.5E-09	1.6E-09
Brain	5.6E-09	1.1E-09	6.3E-10	6.3E-10	6.6E-10	1.0E-10
Breast	3.9E-09	8.7E-10	4.2E-10	3.7E-10	3.0E-10	7.5E-11
GI-Tract						
St Wall	9.6E-09	4.4E-09	2.2E-09	1.4E-09	1.0E-09	6.3E-10
SI Wall	1.3E-08	9.6E-09	5.2E-09	3.4E-09	2.2E-09	1.7E-09
ULI Wall	7.2E-08	6.0E-08	3.0E-08	1.8E-08	9.6E-09	8.5E-09
LLI Wall	2.2E-07	1.9E-07	9.9E-08	5.7E-08	3.1E-08	2.9E-08
Kidneys	6.8E-09	1.7E-09	9.0E-10	7.8E-10	7.2E-10	2.3E-10
Liver	4.9E-09	1.4E-09	7.1E-10	5.5E-10	4.7E-10	1.5E-10
Lungs	5.7E-09	1.2E-09	5.8E-10	5.2E-10	5.1E-10	1.1E-10
Muscle	6.0E-09	1.6E-09	8.5E-10	7.1E-10	6.4E-10	2.0E-10
Ovaries	9.9E-09	5.8E-09	3.3E-09	2.4E-09	1.8E-09	1.1E-09
Pancreas	5.5E-09	1.5E-09	7.9E-10	6.6E-10	5.9E-10	1.9E-10
Red Marrow	7.7E-08	1.2E-08	5.5E-09	5.4E-09	5.5E-09	1.2E-09
Skin	4.8E-09	1.1E-09	5.6E-10	4.7E-10	4.2E-10	1.1E-10
Spleen	5.5E-09	1.4E-09	7.3E-10	5.8E-10	4.9E-10	1.6E-10
Testes	4.1E-09	1.4E-09	7.6E-10	5.5E-10	4.3E-10	1.6E-10
Thymus	4.0E-09	8.9E-10	4.5E-10	4.0E-10	3.9E-10	8.9E-11
Thyroid	3.7E-09	8.4E-10	4.6E-10	4.3E-10	4.6E-10	8.7E-11
Uterus	6.2E-09	3.1E-09	1.7E-09	1.2E-09	9.2E-10	5.0E-10
Remainder	6.0E-09	1.7E-09	9.1E-10	7.6E-10	6.7E-10	2.3E-10
Effective Dose	3.2E-08	1.8E-08	9.2E-09	5.8E-09	3.7E-09	2.6E-09

GI-Tract Gastrointestinal Tract
St Stomach
SI Small Intestine
ULI Upper Large Intestine
LLI Lower Large Intestine

* In the biokinetic model for Ba parameter values for the adult apply to ages > 25 y. For radioisotopes of this element the dose coefficients for the adult are based on the 50-y integrated doses following an acute intake at age 25 y.

References

Bauer, G. C. H., Carlsson, A. and Lindquist, B. (1957) Metabolism of Ba-140 in man. *Acta Orth. Scand.* **26**, 241–257.

Bligh, P. H. (1960) Metabolism of barium in rat and man. PhD Thesis, University of London, England.

Boender, C. A. and Verloop, M. C. (1969) Iron absorption, iron loss and iron retention in man: Studies after oral administration of a tracer dose of $^{59}FeSO_4$ and $^{131}BaSO_4$. *Br. J. Haematol.* **17**, 45–58.

Cuddihy, R. G. and Griffith, W. C. (1972) A biological model describing tissue distribution and whole-body retention of barium and lanthanum in beagle dogs after inhalation and gavage. *Health Phys.* **23**, 621–633.

Della Rosa, R. J., Goldman, M. and Wolf, H. G. (1967) Uptake and retention of orally administered Ra-226 and Ba-133: A preliminary report. University of California at Davis, **472–114**, 40–41.

Domanski, T., Witkowska, D. and Garlicka, I. (1980) Influence of age on the discrimination of barium in comparison with strontium during their incorporation into compact bone. *Acta Physiol. Pol.* **31**, 289–296.

Ellsasser, J. C., Farnham, J. E. and Marshall, J. H. (1969) Comparative kinetics and autoradiography of Ca-45 and Ba-133 in ten-year-old beagle dogs. *J. Bone Jt. Surg.* **51A**, 1397–1412.

Erre, N., Manca, F. and Parodo, A. (1980) The short-term retention of barium in man. *Health Phys.* **38**, 225–227.

Figueroa, W. G., Jordan, T. and Basset, S. H. (1968) Use of barium sulfate as an unabsorbable fecal marker. *Am. J. Clin. Nutr.* **21**, 1239–1245.

Garner, R. J., Jones, H. G. and Sansom, B. F. (1960) Fission products and the dairy cow. *Biochem. J.* **76**, 572–579.

Hardy, E., Rivera, J., Fisenne, I., Pond, W. and Hogue, D. (1969) Comparative utilization of dietary radium-226 and other alkaline earths by pigs and sheep. In: *Radiation Biology of the Fetal and Juvenile Mammal* (M. R. Sikov and D. D. Mahlhum, Eds), pp. 183–190. Proceedings of the Ninth Annual Hanford Biology Symposium at Richland, Washington, May 5–8, 1969; U.S. AEC, Division of Technical Information.

Harrison, G. E. (1981) Whole body retention of the alkaline earths in adult man. *Health Phys.* **40**, 95–99.

Harrison, G. E., Carr, T. E. F. and Sutton, A. (1967) Distribution of radioactive calcium, strontium, barium and radium following intravenous injection into a healthy man. *Int. J. Radiat. Biol.* **13**, 235–247.

ICRP (1973) *Alkaline Earth Metabolism in Adult Man*, ICRP Publication 20. Pergamon Press, Oxford.

ICRP (1979) *Limits for Intake of Radionuclides by Workers: Part 1*, ICRP Publication 30. Pergamon Press, Oxford.

Korsunskii, V. N., Tarasov, N. F. and Naumenko, A. Z. (1981) Clinical evaluation of Ba-133m as an osteotropic agent. ORNL/TR-86/30 (7pgs). Translated from the Russian UDC 616.71-006-073.916. *Medit. Radiol.* **10**, 45–48.

Leggett, R. W. (1992) A generic age-specific biokinetic model for calcium-like elements. *Radiat. Prot. Dosim.* **41**, 183–198.

LeRoy, G. V., Rust, J. H. and Hasterlik, R. J. (1966) The consequences of ingestion by man of real and simulated fallout. *Health Phys.* **12**, 449–473.

van Middlesworth, L. and Robinson, W. L. (1975) Thyroid concentration of barium and radium. *Int. J. Nucl. Med. Biol.* **2**, 1–4.

Newton, D., Rundo, J. and Harrison, G. E. (1977) The retention of alkaline earth elements in man, with special reference to barium. *Health Phys.* **33**, 45–53.

Newton, D., Harrison, G. E., Kang, C. and Warner, A. J. (1991) The Metabolism of injected barium in six healthy men. *Health Phys.* **61**, 191–201.

Rundo, J. (1967) The retention of barium-133 in man. *Int. J. Radiat. Biol.* **13**, 301–302.

Sansom, B. F. and Garner, R. J. (1966) The metabolism of radium in dairy cows. *Biochem. J.* **99**, 677–681.

Schroeder, H. A., Tipton, I. H. and Nason, A. P. (1972) Trace metals in man: strontium and barium. *J. Chron. Dis.* **25**, 491–517.

Stather, J. W. (1974) Distribution of P-32, Ca-45, Sr-85 and Ba-133 as a function of age in the mouse skeleton. *Health Phys.* **26**, 71–79.

Taylor, D. M., Bligh, P. H. and Duggan, M. H. (1962) The absorption of calcium, strontium, barium, and radium from the gastrointestinal tract of the rat. *Biochem. J.* **83**, 25–29.

Tipton, I. H., Stewart, P. L. and Dickson, J. (1969) Patterns of elemental excretion in long term balance studies. *Health. Phys.* **16**, 455–462.

11. LEAD

Uptake to blood

(104) *Adults.* The absorption of lead (Pb) in man has been studied extensively (ICRP, 1975; Moore, 1986; NEA/OECD, 1988). In normally fed adults, f_1 values range from 0.03 to 0.14. Lead is absorbed more readily in fasting subjects than when it is ingested with food (range of f_1 values 0.1–0.76). ICRP (1980) recommended an f_1 value of 0.2 for lead for occupationally exposed adults. This is also a reasonable value to apply to adult members of the public and is adopted here.

(105) *Children.* Ziegler *et al.* (1978) reported repeated 3-day balance studies of lead in six normal infants of ages ranging from 14 to 746 days, most being in the range 150–300 days. In subjects ingesting more than 5 μg of lead kg^{-1} body weight day^{-1}, the net absorption averaged 42%. In eight children of ages 3 months to 8 years with daily intakes of 5–17 μg kg^{-1} body weight, Alexander *et al.* (1974) reported absorption of 53% and retention of 18%. There are a number of uncertainties in these estimates and true absorption could have been greater due to endogenous faecal excretion of absorbed lead. However, the results suggest that the fractional absorption of lead in children may remain high for at least the first 8 years of life. Results for non-human primates are qualitatively consistent with these data for humans (Willes *et al.*, 1977; Pounds *et al.*, 1978). Studies in chickens, rats and sheep demonstrate clearly that there is an inverse relationship of age and the capacity to absorb lead (Kostial *et al.*, 1974; Gerber and DeRoo, 1974; Conrad and Barton, 1978; Quaterman and Morrison, 1978; Mykkänen and Wasserman, 1981). An f_1 value of 0.6 is adopted here for the 3-month-old infant. For ages 1–15 years a value of 0.4 is used.

Distribution and retention

(106) *Adults.* At 24 h after intravenous injection of radioactive lead into adult humans or baboons, the blood, liver and skeleton contained about 55%, 15% and 10–15% of the administered activity, respectively (Hursh and Suomela, 1968; Booker *et al.*, 1969; Hursch and Brown, 1969; Cohen *et al.*, 1970; Wells *et al.*, 1975; Chamberlain *et al.*, 1978; Morrow *et al.*, 1980; Heard and Chamberlain, 1984). Most of the remainder is associated with other soft tissues, but 3–6% may be excreted during the first day (Hursh and Suomela, 1968; Hursh *et al.*, 1969; Booker *et al.*, 1969; Hursh and Mercer, 1970). More than 99% of the activity in blood is associated with red blood cells (Chamberlain *et al.*, 1978; DeSilva, 1981), which lose about half of their lead content within 15–20 days (Rabinowitz *et al.*, 1973, 1974, 1976; Wells *et al.*, 1975; Heard and Chamberlain, 1984). Injection studies in humans (Heard and Chamberlain, 1984), baboons (Cohen *et al.*, 1970) and beagles (Lloyd *et al.*, 1975) indicate that lead is lost from the liver with an apparent biological half-time of a few weeks. Gradual loss of lead from red blood cells, liver and other soft tissues over the first few weeks can be accounted for by a slow loss in urine and faeces (about 30% of the systemic activity in the first 20 days) and a continual increase in skeletal lead (Heard and Chamberlain, 1984). By 20 days after injection of lead, the human skeleton may contain about 25% of the injected amount (Heard and Chamberlain, 1984). Autopsy data for persons chronically exposed to environmental lead indicate that the skeletal content of lead increases throughout life and may represent

90% or more of systemic lead by the fifth decade (Tipton and Cook, 1963; Gross *et al.*, 1975; Barry, 1975, 1981). Skeletal behaviour of lead appears to be similar to that of the heavier alkaline earth elements, if account is taken of the slower deposition of lead in the skeleton due to competition with other tissues and fluids, particularly red blood cells (Hursh, 1973; Lloyd *et al.*, 1975; Domanski and Trojanowska, 1980; Heard and Chamberlain, 1984).

(107) The age-specific model (Leggett, 1992, 1993) adopted here for lead is similar in structure to the models for barium and radium, but account is taken of the considerable differences in the time-dependent distribution of lead and the alkaline earths and in their rates of excretion. The model for lead is described in Appendix A.

(108) *Children.* Autopsy data for persons of all ages exposed only to environmental lead indicate that the bones of infants may contain roughly one-quarter to one-third of total-body lead and those of young children and teenagers one-half to two-thirds, compared with a value of 90% or more for middle-aged and older persons (Barry, 1973, 1975, 1981; Leggett, 1992). Data for non-human primates (Willes *et al.*, 1977; Pounds *et al.*, 1978; Kneip *et al.*, 1983) and rodents (Momcilovic and Kostial, 1974; Kello *et al.*, 1975; Jugo *et al.*, 1975; Kostial *et al.*, 1978; Jugo, 1980; Keller and Doherty, 1980) indicate that retention of lead is greater in growing than in mature animals and that much of the variation with age is due to elevated uptake and/or retention of lead by the immature skeleton. Some differences in uptake and/or retention of lead by the brain, liver and kidneys have been observed between immature and mature rodents (Momcilovic and Kostial, 1974; Keller and Doherty, 1980). The combined age-specific data for laboratory animals and environmentally exposed humans are reasonably consistent with assumptions of higher uptake and faster release of lead by the immature than the mature skeleton and substantial retention of lead by blood and soft tissues at all ages. The selection of age-specific biokinetic parameters for the model adopted here for lead is described in Appendix A.

Behaviour of daughters

(109) Radioactive progeny of lead radioisotopes produced in bone volume are assumed to follow the behaviour of the parent radionuclide until removed from the bone volume. Outside bone volume, retention and translocation of decay products of lead radioisotopes are described by a separate characteristic biokinetic model for each element rather than the biokinetic model for the absorbed lead radioisotope.

(110) For polonium produced *in vivo*, removal from all tissues (other than bone volume) to plasma is assumed to occur at a rate of 0.1 day^{-1}. Of polonium reaching plasma, 10% goes to the gastrointestinal tract contents and subsequently to faeces, 5% goes to urine, and the remaining 85% is distributed as indicated in Table 12.-1. The removal rate from plasma is 100 day^{-1}. This model for polonium as a daughter yields integrated activities in organs similar to those produced by the catenary model for polonium described in Chapter 12.

(111) A substantial portion of bismuth reaching the systemic circulation is rapidly excreted in urine (Coenegracht and Dorleyn, 1961; Matthews *et al.*, 1964; Hursh and Brown, 1969; Russ *et al.*, 1975; Pieri and Wegmann, 1981; Vienet *et al.*, 1983; Gregus and Klaassen, 1986), but some is retained for a period of weeks (Sollmann, 1957; Boiteau *et al.*, 1976; Loiseau *et al.*, 1976; Froomes *et al.*, 1989) and a small portion may be retained for many years (Buijs *et al.*, 1985). This is in contrast to the relatively rapid removal of bismuth from the body assumed in *ICRP Publication 30* (ICRP, 1981). Since

the bismuth isotopes of concern here have radiological half-lives of at most a few days, extremely long-term retention would be of little consequence for the present applications and is not addressed in the model for bismuth as a daughter. It is assumed that bismuth produced *in vivo* is removed from all tissues (other than bone volume) to plasma at a rate of 0.035 day^{-1}. Of bismuth reaching plasma, 35% goes to urine, 7% to the gastrointestinal tract contents and subsequently to faeces, 35% to the kidneys, 5% to the liver and 18% to other tissues (Sollmann, 1957; Coenegracht and Dorleyn, 1961; Russ *et al.*, 1975; Gregus and Klaassen, 1986). The rate of removal from plasma is 50 day^{-1}.

Dose coefficients

(112) Dose coefficients derived from the biokinetic data summarized in Table A-4 are given in Table 11.-1.

Table 11.-1

Ingestion Dose Coefficients (Equivalent Dose (Sv/Bq) to Age 70 Years) for Pb-210 (T1/2 = 22.3 y)*

Age at intake	3 Months	1 Year	5 Years	10 Years	15 Years	Adult
Adrenals	2.1E-06	9.6E-07	5.3E-07	3.9E-07	2.7E-07	9.1E-08
Bladder Wall	2.1E-06	9.6E-07	5.3E-07	3.9E-07	2.8E-07	9.2E-08
Bone Surfaces	8.6E-05	3.8E-05	3.3E-05	4.5E-05	7.0E-05	2.3E-05
Brain	2.1E-06	9.6E-07	5.3E-07	3.9E-07	2.7E-07	9.1E-08
Breast	2.1E-06	9.6E-07	5.3E-07	3.9E-07	2.7E-07	9.1E-08
GI-Tract						
St Wall	2.1E-06	9.6E-07	5.3E-07	3.9E-07	2.7E-07	9.1E-08
SI Wall	2.1E-06	9.6E-07	5.3E-07	3.9E-07	2.7E-07	9.1E-08
ULI Wall	2.2E-06	9.9E-07	5.5E-07	4.0E-07	2.8E-07	9.4E-08
LLI Wall	2.3E-06	1.1E-06	5.8E-07	4.2E-07	3.0E-07	1.0E-07
Kidneys	6.3E-05	2.8E-05	1.7E-05	1.5E-05	1.2E-05	3.7E-06
Liver	3.7E-05	1.7E-05	9.7E-06	8.3E-06	6.0E-06	1.9E-06
Lungs	2.1E-06	9.6E-07	5.3E-07	3.9E-07	2.7E-07	9.1E-08
Muscle	2.1E-06	9.6E-07	5.3E-07	3.9E-07	2.7E-07	9.1E-08
Ovaries	2.1E-06	9.6E-07	5.3E-07	3.9E-07	2.7E-07	9.1E-08
Pancreas	2.1E-06	9.6E-07	5.3E-07	3.9E-07	2.7E-07	9.1E-08
Red Marrow	3.0E-05	1.3E-05	7.2E-06	6.1E-06	5.7E-06	2.4E-06
Skin	2.1E-06	9.6E-07	5.3E-07	3.9E-07	2.7E-07	9.1E-08
Spleen	6.7E-05	3.0E-05	1.7E-05	1.4E-05	9.7E-06	2.7E-06
Testes	2.1E-06	9.6E-07	5.3E-07	3.9E-07	2.7E-07	9.1E-08
Thymus	2.1E-06	9.6E-07	5.3E-07	3.9E-07	2.7E-07	9.1E-08
Thyroid	2.1E-06	9.6E-07	5.3E-07	3.9E-07	2.7E-07	9.1E-08
Uterus	2.1E-06	9.6E-07	5.3E-07	3.9E-07	2.7E-07	9.1E-08
Remainder	3.6E-06	1.6E-06	9.1E-07	6.3E-07	4.4E-07	1.4E-07
Effective Dose	8.1E-06	3.6E-06	2.1E-06	1.9E-06	1.9E-06	7.0E-07

GI-Tract Gastrointestinal Tract
St Stomach
SI Small Intestine
ULI Upper Large Intestine
LLI Lower Large Intestine

* In the biokinetic model for Pb parameter values for the adult apply to ages > 25 y. For radioisotopes of this element the dose coefficients for the adult are based on the 50-y integrated doses following an acute intake at age 25 y.

References

Alexander, F. W., Clayton, B. E. and Delves, H. T. (1974) Mineral and trace metal balances in children receiving normal and synthetic drugs. *Q. J. Med.* **43**, 89–111.

Barry, P. S. I. (1973) A comparison of lead concentrations in human bones and in soft tissues. In: *Environmental Health Aspects of Lead*, Proceedings of Int. Symp. of the Commission of the European Communities and the US EPA, Amsterdam, October 2–6, 1972, Luxembourg, pp. 415–426.

Barry, P. S. I. (1975) A comparison of lead concentrations in human tissues. *Br. J. Ind. Med.* **32**, 119–139.

Barry, P. S. I. (1981) Concentrations of lead in tissues of children. *Br. J. Ind. Med.* **38**, 61–71.

Boiteau, H. L., Cler, J. M., Mathé, J. F., Delobel, R. and Fève, J. R. (1976) Relations entre l'évolution des encéphalopathies bismuthiques et les taux de bismuth dans le sang et dans les urines. *Eur. J. Toxicol.* **9**, 233–239.

Booker, D. V., Chamberlain, A. C., Newton, D. and Stott, A. N. B. (1969) Uptake of radioactive lead following inhalation and injection. *Br. J. Radiol.* **42**, 457–466.

Buijs, W. C., Corstens, F. H. and Beentjes, L. B. (1985) Long term retention of ^{207}Bi in the human body after injection of ^{206}Bi. *Health Phys.* **49**, 1267–1269.

Chamberlain, A. C., Heard, M. J., Little, P., Newton, D., Wells, A. C. and Wiffen, R. D. (1978) *Investigations into Lead from Motor Vehicles.* AERE-R 9198. AERE, Environmental and Medical Sciences Division, Harwell.

Coenegracht, J. M. and Dorleyn, M. (1961) The distribution of intravenously administered tracer doses of Bi-206 compounds in the human body. *J. Belg. Radiol.* **44**, 485–504.

Cohen, N., Eisenbud, M. and Wrenn, M. E. (1970) *The Retention and Distribution of Lead-210 in the Adult Baboon.* Progress report, Radioactivity studies, New York University Medical Center, New York.

Conrad, M. E. and Barton, J. C. (1978) Factors affecting the absorption and excretion of lead in the rat. *Gastroenterology* **74**, 731–740.

DeSilva, P. E. (1981) Determination of lead in plasma and studies on its relationship to lead in erythrocytes. *Br. J. Ind. Med.* **38**, 209–217.

Domanski, T. and Trojanowska, B. (1980) Studies on metabolic kinetics of lead and alkaline earth elements (Ca, Ba). *Acta Physiol. Pol.* **31**, 439–447.

Froomes, P. R. A., Wan, A. T., Keech, A. C., McNeil, J. J. and McLean, A. J. (1989) Absorption and elimination of bismuth from oral doses of tripotassium dicitrato bismuthate. *Eur. J. Pharmacol.* **37**, 533–536.

Gerber, G. B. and DeRoo, J. (1974) Absorption of radioactive lead (^{210}Pb) by different parts of the intestine in young and adult rats. *Environ. Physiol. Biochem.* **5**, 314–318.

Gregus, Z. and Klaassen, C. D. (1986) Disposition of metals in rats: a comparative study of fecal, urinary, and biliary excretion and tissue distribution of eighteen metals. *Toxicol. Appl. Pharmacol.* **85**, 24–38.

Gross, S. B., Pfitzer, E. A., Yeager, D. W. and Kehoe, R. A. (1975) Lead in human tissues. *Toxicol. Appl. Pharmacol.* **32**, 638–651.

Heard, M. J. and Chamberlain, A. C. (1984) Uptake of Pb by human skeleton and comparative metabolism of Pb and alkaline earth elements. *Health Phys.* **47**, 857–865.

Hursh, J. B. (1973) Retention of Pb-210 in beagle dogs. *Health Phys.* **25**, 29–35.

Hursh, J. B. and Brown, C. (1969) Tissue distribution of ^{212}Bi in rats. *Proc. Soc. Exp. Biol. Med.* **131**, 116–120.

Hursh, J. B. and Mercer, T. T. (1970) Measurement of Pb-212 loss rate from human lungs. *J. Appl. Physiol.* **28**(3), 268–274.

Hursh, J. B. and Suomela, J. (1968) Absorption of Pb-212 from the gastrointestinal tract of man. *Acta Radiol., Ther. Phys. Biol.* **7**, 108–120.

Hursh, J. B., Schraub, A., Sattler, E. L. and Hofmann, H. P. (1969) Fate of Pb-212 inhaled by human subjects. *Health Phys.* **16**, 257–267.

ICRP (1975) *Report of the Task Group on Reference Man,* ICRP Publication 23, pp. 390–391. Pergamon Press, Oxford.

ICRP (1980) *Limits for Intakes of Radionuclides by Workers: Part 2,* ICRP Publication 30, pp. 64–65. Pergamon Press, Oxford.

ICRP (1981) *Limits for Intakes of Radionuclides by Workers: Part 3,* ICRP Publication 30. Pergamon Press, Oxford.

Jugo, S. (1980) Chelatable fraction of Pb-203 in blood of young and adult rats. *Environ. Res.* **21**, 336–342.

Jugo, S., Maljkovic, T. and Kostial, K. (1975) The effect of chelating agents on lead excretion in rats in relation to age. *Environ. Res.* **10**, 271–279.

Keller, C. A. and Doherty, R. A. (1980) Distribution and excretion of lead in young and adult female mice. *Environ. Res.* **21**, 217–228.

Kello, D., Kostial, K. and Harrison, G. E. (1975) Influence of age and temperature of incineration on the loss of lead from rat bone. *Health Phys.* **28**, 169–170.

Kneip, T. J., Mallon, R. P. and Harley, N. H. (1983) Biokinetic modelling for mammalian lead metabolism. *Neurotoxicology* **4**(3), 189–192.

Kostial, K., Malijkovic, T. and Jugo, S. (1974) Lead acetate toxicity in rats in relation to age and sex. *Arch. Toxicol.* **31**, 265–269.

Kostial, K., Kello, D., Jugo, S., Rabar, I. and Maljkovic, T. (1978) Influence of age on metal metabolism and toxicity. *Environ. Health Perspect.* **25**, 81–86.

Leggett, R. W. (1992) A generic age-specific biokinetic model for calcium-like elements. Paper presented at the Workshop on "Age-dependent Factors in the Biokinetics and Dosimetry of Radionuclides", Schloß Elmau, Germany, 5–8 November 1991. *Radiat. Prot. Dosim.* **41**, 183–198.

Leggett, R. W. (1993) An age-specific kinetic model of lead in humans. *Environ. Health Perspect.* **101**, 598–616.

Lloyd, R. D., Mays, C. W., Atherton, D. R. and Bruenger, F. W. (1975) ^{210}Pb studies in beagles. *Health Phys.* **28**, 575–583.

Loiseau, P., Henry, P., Jallon, P. and Legroux, M. (1976) Encéphalopathies myocloniques iatrogènes aux sels de bismuth. *J. Neurol. Sci.* **27**, 133–143.

Matthews, C. M. E., Dempster, W. J., Kapros, C. and Kountz, S. (1964) The effect of bismuth 206 irradiation on survival of skin homografts. *Br. J. Radiol.* **37**, 306–310.

Momcilovic, B. and Kostial, K. (1974) Kinetics of lead retention and distribution in suckling and adult rats. *Environ. Res.* **8**, 214–220.

Moore, M. R. (1986) Lead in humans. In: *Lead Toxicity. History and Environmental Impact* (R. Lansdown and W. Yule, Eds), pp. 54–95. Johns Hopkins University Press, Baltimore.

Morrow, P. E., Beiter, H., Amato, F. and Gibb, F. R. (1980) Pulmonary retention of lead: An experimental study in man. *Environ. Res.* **21**, 373–384.

Mykkänen, H. and Wasserman, R. H. (1981) Gastrointestinal absorption of lead (^{203}Pb) in chicks: influence of lead, calcium and age. *J. Nutr.* **111**, 1757–1765.

NEA/OECD (1988) Committee on Radiation Protection and Public Health. *Report of an Expert Group on Gut Transfer Factors*, pp. 45–60. NEA/OECD Report, Paris.

Pieri, F. and Wegmann, R. (1981) Radiopharmaceutical study of Bi-citrate in the rat. Compartmental distribution. *Cell. Molec. Biol.* **27**, 57–60.

Pounds, J. G., Marlar, R. J. and Allen, J. R. (1978) Metabolism of lead-210 in juvenile and adult rhesus monkeys (Macaca mulatta). *Bull. Environ. Contam. Toxicol.* **19**, 684–691.

Quaterman, J. and Morrison, E. (1978) The effect of age on the absorption and excretion of lead. *Environ. Res.* **17**, 78–83.

Rabinowitz, M. B., Wetherill, G. W. and Kopple, J. D. (1973) Lead metabolism in the normal human: Stable isotope studies. *Science* **182**, 725–727.

Rabinowitz, M. B., Wetherill, G. W. and Kopple, J. D. (1974) Studies of human lead metabolism by use of stable isotope tracers. *Environ. Health Perspect.* **5–7**, 145–153.

Rabinowitz, M. B., Wetherill, G. W. and Kopple, J. D. (1976) Kinetic analysis of lead metabolism in healthy humans. *J. Clin. Invest.* **58**, 260–270.

Russ, G. A., Bigler, R. E., Tilbury, R. S., Woodard, H. Q. and Laughlin, J. S. (1975) Metabolic studies with radiobismuth. I. Retention and distribution of ^{206}Bi in the normal rat. *Radiat. Res.* **63**, 443–454.

Sollmann, T. A. (1957) *Manual of Pharmacology and its Applications to Therapeutics and Toxicology.* W. B. Saunders, Philadelphia.

Tipton, I. H. and Cook, M. J. (1963) Trace elements in human tissue. Part II. Adult subjects from the United States. *Health Phys.* **9**, 103–145.

Vienet, R., Bouvet, P. and Istin, M. (1983) Kinetics and distribution of ^{206}Bi in the rat and the rabbit: a model. *Int. J. Appl. Radiat. Isot.* **34**, 747–753.

Wells, A. C., Venn, J. B. and Heard, M. J. (1975) Deposition in the lung and uptake to blood of motor exhaust labelled with Pb-203. In: *Inhaled Particles IV* (W. H. Walton and B. McGovern, Eds), pp. 175–189. Proceedings of a Symposium of the British Occupational Hygiene Society, Edinburgh.

Willes, R. F., Lok, E., Truelove, J. F. and Sundaram, A. (1977) Retention and tissue distribution of Pb-210 $(NO_3)_2$ administered orally to infant and adult monkeys. *J. Toxicol. Environ. Health* **3**, 395–406.

Ziegler, E. E., Edwards, B. B., Jensen, R. L., Mahaffey, K. R. and Fomon, S. J. (1978) Absorption and retention of lead by infants. *Pediat. Res.* **12**, 29–34.

12. POLONIUM

Uptake to blood

(113) *Adults.* Few data are available regarding the fractional absorption of inorganic polonium (Po) in man. An f_1 value of approximately 0.1 was determined in a single male with chronic myelogenous leukaemia (Fink, 1950). In rats, fractional absorption values of 0.03–0.06 were determined using an unspecified polonium compound (Anthony *et al.*, 1956). Della Rosa *et al.* (1955) measured fractional absorption of 0.06 in mature rats given polonium chloride by gavage. Similarly, f_1 values of 0.04–0.06 for polonium as the nitrate and 0.06–0.08 for the citrate were obtained (Naylor *et al.*, 1991). Data from humans consuming meat from reindeer exposed to ^{210}Po suggest that polonium incorporated in food may have an f_1 of 5–25 times that of inorganic polonium compounds, values of around 0.3 to 0.5 having been estimated (Hill, 1965; Kauranen and Miettinen, 1967; Landinskaya *et al.*, 1973). Rats absorbed approximately 50% of polonium biologically incorporated into goat's milk. This value is higher than is typically observed for inorganic polonium absorption in rats (McInroy *et al.*, 1972). Similarly, experiments in rats ingesting polonium biologically incorporated into rat liver yielded values of f_1 of 0.10–0.12 which are at least twice those which have been obtained using inorganic polonium compounds (Naylor *et al.*, 1991). In a recent human volunteer study of the absorption of ^{210}Po from crabmeat, Hunt *et al.* (1993) estimated absorption to be about 0.8. ICRP (1979) previously recommended an f_1 value of 0.1 for workers. In the present report, an f_1 value of 0.5 has been adopted for dietary intakes of Po by adult members of the public.

(114) *Children.* There appear to be no data available for estimating an f_1 value for infants and children. Therefore, following the general approach of the NEA/OECD expert group (NEA/OECD, 1988), an f_1 value of 1.0 for polonium is adopted here for the 3-month-old infant. For children of 1 year and older the f_1 value for the adult (0.5) is used here.

Distribution and retention

(115) *Adults.* Polonium entering the bloodstream is deposited predominantly in soft tissues. In rats injected intravenously with polonium, the concentration in the reticuloendothelial system is approximately 10 times that in other soft tissues, with lower concentrations in the muscle and cortical bone (Stannard and Cassarett, 1964). In studies of beagle dogs exposed to an aerosol containing ^{210}Po the activity translocated to the blood was deposited mainly in the liver, kidneys and spleen (Smith *et al.*, 1961). In humans, highest concentrations have been found in liver, lung, kidney and rib with respect to intakes of naturally occurring ^{210}Po (Chen, 1987). In a study on rats values of 11%, 7% and 5% were obtained for deposition in the liver, kidneys and spleen, respectively, 10 days after intravenous injection of ^{210}Po as the chloride (Fink, 1950). Fink also obtained values for a single human suffering from acute leukaemia at 6 days after injection of ^{210}Po as the chloride, which showed approximately 40%, 5% and 4% of injected activity retained in the liver, kidneys and spleen, respectively. Naylor *et al.* (1991) have reported that in marmosets the highest deposit of ^{210}Po was in the liver (26% of the total body activity) at 1 week after i.v. injection as the citrate, with appreciable deposits in the kidney (21%) and much less in the spleen and testes (1% and

0.5%, respectively). The femora contained 1.5% of the retained activity, which corresponds to a skeletal deposit of about 15% of the total systemic deposit. At 4 weeks the tissue distribution was similar to that found at 7 days, although the femoral activity had fallen to 0.75% of the total body activity. In rats, retention of polonium in the liver, kidney, spleen and femora at 1 week after i.v. injection of ^{210}Po as the citrate has been reported to be 7.8%, 5.3%, 7.3% and 0.69%, respectively (Naylor *et al.*, 1991).

(116) More than 50% of polonium entering the blood is excreted in faeces (Fink, 1950; Stannard and Cassarett, 1964; Fellman *et al.*, 1989). Whole-body retention half-times for polonium have been estimated to be 30 days (ICRP, 1960) and 50 days (Jackson and Dolphin, 1964; ICRP, 1968, 1979). The biokinetic model for polonium in *ICRP Publication 30* (ICRP, 1979) assumed that of polonium entering the blood, fractions 0.1, 0.1, 0.1 and 0.7 go to the liver, kidneys, spleen and all other tissues, respectively, and are retained there with a half-time of 50 days. The more recent data discussed above suggest that it is appropriate to take account of deposits in the bone marrow in assessing tissue doses. Recent autoradiographic studies also indicate that virtually all skeletal polonium is retained in the marrow (Naylor *et al.*, 1991). It is therefore assumed that all polonium retained in the skeleton is associated with the red bone marrow. For the purposes of dose calculations, it has therefore been assumed that of polonium entering the systemic circulation the fractions 30%, 10%, 5%, 10% and 45% are deposited in the liver, kidneys, spleen, red bone marrow and the rest of the body, respectively, where it is retained with a half-time of 50 days. A urinary to faecal excretion ratio of 1:2 is assumed for polonium that has entered the transfer compartment.

(117) *Children.* Two studies involving a total of 14 children and adolescents aged between 6 and 15 years exposed to ^{210}Po indicated a half-time of retention of about 40 days (Guskova *et al.*, 1964; Kalmykov *et al.*, 1969). This value is not significantly different from the measured values for adults. The distribution and retention of polonium are assumed to be independent of age.

Dose coefficients

(118) Dose coefficients derived from the biokinetic data summarized in Table 12.-1 are given in Table 12.-2.

Table 12.-1. Biokinetic data for polonium

		Distribution (%)					Biological half-time (days)
Age	f_1	Liver	Kidney	Spleen	Red bone marrow	Other tissues	All tissues
3 months	1.0	30	10	5	10	45	50
1 year	0.5	30	10	5	10	45	50
5 years	0.5	30	10	5	10	45	50
10 years	0.5	30	10	5	10	45	50
15 years	0.5	30	10	5	10	45	50
Adult	0.5	30	10	5	10	45	50

A urinary to faecal excretion ratio of 1:2 is assumed for polonium that has entered the transfer compartment.

Table 12.-2

Ingestion Dose Coefficients (Equivalent Dose (Sv/Bq) to Age 70 Years) for Po-210 (T1/2 = 138.38 d)

Age at intake	3 Months	1 Year	5 Years	10 Years	15 Years	Adult
Adrenals	4.6E-06	2.0E-06	9.8E-07	5.8E-07	3.4E-07	2.8E-07
Bladder Wall	4.6E-06	2.0E-06	9.8E-07	5.8E-07	3.4E-07	2.8E-07
Bone Surfaces	6.4E-05	2.3E-05	8.9E-06	4.7E-06	2.8E-06	1.6E-06
Brain	4.6E-06	2.0E-06	9.8E-07	5.8E-07	3.4E-07	2.8E-07
Breast	4.6E-06	2.0E-06	9.8E-07	5.8E-07	3.4E-07	2.8E-07
GI-Tract						
St Wall	4.6E-06	2.0E-06	9.8E-07	5.8E-07	3.4E-07	2.8E-07
SI Wall	4.6E-06	2.0E-06	9.8E-07	5.9E-07	3.4E-07	2.8E-07
ULI Wall	4.7E-06	2.1E-06	1.0E-06	6.1E-07	3.6E-07	2.9E-07
LLI Wall	4.9E-06	2.3E-06	1.1E-06	6.7E-07	3.9E-07	3.2E-07
Kidneys	1.4E-04	6.1E-05	3.4E-05	2.3E-05	1.6E-05	1.3E-05
Liver	8.8E-05	4.0E-05	2.0E-05	1.3E-05	8.4E-06	6.6E-06
Lungs	4.6E-06	2.0E-06	9.8E-07	5.8E-07	3.4E-07	2.8E-07
Muscle	4.6E-06	2.0E-06	9.8E-07	5.8E-07	3.4E-07	2.8E-07
Ovaries	4.6E-06	2.0E-06	9.8E-07	5.8E-07	3.4E-07	2.8E-07
Pancreas	4.6E-06	2.0E-06	9.8E-07	5.8E-07	3.4E-07	2.8E-07
Red Marrow	6.5E-05	2.6E-05	1.2E-05	6.4E-06	3.8E-06	2.6E-06
Skin	4.6E-06	2.0E-06	9.8E-07	5.8E-07	3.4E-07	2.8E-07
Spleen	1.8E-04	7.6E-05	4.1E-05	2.5E-05	1.6E-05	1.1E-05
Testes	4.6E-06	2.0E-06	9.8E-07	5.8E-07	3.4E-07	2.8E-07
Thymus	4.6E-06	2.0E-06	9.8E-07	5.8E-07	3.4E-07	2.8E-07
Thyroid	4.6E-06	2.0E-06	9.8E-07	5.8E-07	3.4E-07	2.8E-07
Uterus	4.6E-06	2.0E-06	9.8E-07	5.8E-07	3.4E-07	2.8E-07
Remainder	9.3E-05	3.9E-05	2.1E-05	1.3E-05	8.2E-06	6.6E-06
Effective Dose	2.1E-05	8.8E-06	4.4E-06	2.6E-06	1.6E-06	1.2E-06

GI-Tract	Gastrointestinal Tract
St	Stomach
SI	Small Intestine
ULI	Upper Large Intestine
LLI	Lower Large Intestine

References

Anthony, D. S., Davis, R. K., Cowden, R. N. and Jolley, W. P. (1956) Experimental data useful in establishing maximum single and multiple exposure to polonium. In: *Proceedings of the International Conference on the Peaceful Uses of Atomic Energy*, Vol. 13, pp. 215–218. United Nations, New York.

Chen, X. (1987) Body content of some naturally occurring radioisotopes in Chinese people. *NRPB Bull.* **84**, 13–20.

Della Rosa, R. J., Thomas, R. G. and Stannard, J. N. (1955) University of Rochester, UR-392.

Fellman, A., Ralston, L., Hickman, D., Ayres, L. and Cohen, N. (1989) The importance of acid digestion of urine prior to spontaneous deposition of ^{210}Po. *Health Phys.* **57**, 615–621.

Fink, R. M. (1950) *Biological Studies with Polonium, Radium and Plutonium*. McGraw-Hill, London.

Guskova, A. K., Drutman, R. D., Malysheva, M. S. and Soldatova, V. A. (1964) The assessment of doses and the possibility of clinical recognition of the disease associated with Po^{210} affection. *Med. Radiol.* **9**(8), 51–60.

Hill, C. R. (1965) ^{210}Polonium in man. *Nature* **208**, 423–428.

Hunt, G. J. and Allington, D. J. (1993) Absorption of environmental polonium-210 by the human gut. *J. Radiol. Prot.* **13**(2), 119–126.

ICRP (1960) A Report by Committee 2 of ICRP. *Health Phys.* **3**, 217–220.

ICRP (1968) *Evaluation of Radiation Doses to Body Tissues from Internal Contamination due to Occupational Exposure.* A Report by Committee 4 of ICRP. ICRP Publication 10. Pergamon Press, Oxford.

ICRP (1979) *Limits for Intakes of Radionuclides by Workers: Part 1,* ICRP Publication 30. Pergamon Press, Oxford.

Jackson, S. and Dolphin, G. W. (1966) The estimation of internal radiation dose from metabolic and urinary excretion data in a number of important radionuclides. *Health Phys.* **12**, 481–500.

Kalmykov, L. Z., Krupchitskaya, K. T., Stratienko, U. V. and Rutkevitch, N. Y. (1969) Characteristics of ^{210}Po removal from the human body. *Med. Radiol.* **14**(12), 26–30.

Kauranen, P. and Miettinen, J. K. (1967) ^{210}Po and ^{210}Pb in environmental samples in Finland. In: *Radiological Concentration Processes* (B. Aberg and F. P. Hungate, Eds), pp. 275–280. Proc. Symp. Stockholm, April 25–29, 1966. Pergamon Press, Oxford.

Landinskaya, L. A., Parfenov, Y. D., Popov, D. K. and Federova, A. V. (1973) ^{210}Pb and ^{210}Po content of air, water, foodstuffs, and the human body. *Arch. Environ. Health* **27**, 254–258.

McInroy, J. F., Watters, R. L. and Johnson, J. E. (1972) Polonium-210 absorption in rats: Effects of biological modification. *Nature New Biol.* **236**, 118–120.

Naylor, G. P. L., Bonas, H. E., Haines, J. W., Ham, G. J., Harrison, J. D., Sundaram, S. and Dayan, A. D. (1991). The gastrointestinal absorption and tissue distribution of alpha-emitting actinide isotopes and polonium-210. The British Nuclear Energy Society Conference: Occupational Radiation Protection, Guernsey. Thomas Telford, London, pp. 291–296.

NEA/OECD (1988) Committee on Radiation Protection and Public Health. A Report of an Expert Group on Gut Transfer Factors. NEA/OECD Report, Paris, 47–49.

Smith, F. A., Morrow, P. E. *et al.* (1961) Distribution and excretion studies in dogs exposed to an aerosol containing polonium-210. *Am. Ind. Hyg. Ass. J.* **22**, 201–208.

Stannard, J. N. and Cassarett, G. W. (1964). Metabolism and biological effects of an alpha-emitter, polonium-210. Academic Press, New York.

13. RADIUM

Uptake to blood

(119) *Adults.* The Task Group on Alkaline Earth Metabolism in Adult Man (ICRP, 1973) estimated that gastrointestinal absorption of radium (Ra) may be about 15–21% of the ingested amount, based on data for whole body ^{226}Ra in adult humans who drank water high in ^{226}Ra for extended periods or ingested ^{226}Ra incorporated in food (Stehney and Lucas, 1956). Normal elderly human subjects ingesting mock radium dial paint containing $^{224}RaSO_4$ absorbed about 20% of ingested radium as an average (Maletskos *et al.*, 1966, 1969). In a study in which an adult male human took 0.05 mg radium by mouth on two occasions, an estimated 25–35% of the ingested amount remained in the body at 5–6 days (Seil *et al.*, 1915). In rats, fasting for 18 h increased absorption of radium in young adult and elderly animals (Taylor *et al.*, 1962).

(120) In *ICRP Publication 30* (ICRP, 1979), an f_1 value of 0.2 was chosen, based mainly on the estimates given in *ICRP Publication 20* (ICRP, 1973). An Expert Group of the NEA/OECD (1988) adopted this value for adults and it is also used here.

(121) *Children.* There is considerable evidence of elevated gastrointestinal absorption of the alkaline earth elements by both laboratory animals and humans during periods of rapid growth, but there is relatively little radium-specific information. Results reported by Muth and Glöbel (1983) for ^{226}Ra in food, water and human bone suggest a much higher rate of transfer of ^{226}Ra from diet to bone during periods of rapid growth than during adulthood or periods of relatively slow growth during childhood. In a study of variation with age in absorption of alkaline earths from the gastrointestinal tract of the rat, Taylor *et al.* (1962) determined that absorption of radium averaged 79% in suckling rats, 11% in young adult rats and 3% in old rats. Similarly, data of Della Rosa *et al.* (1967) suggest considerably greater gastrointestinal absorption of radium in immature than in mature beagles. The available data for radium and other alkaline earth elements, mainly strontium, suggest that two controlling factors may be variation with age in gastrointestinal absorption and corresponding changes in deposition and retention in the skeleton. In the absence of more specific information a fractional absorption of radium of 0.6 is used for the infant. For ages 1–15 years, a value of 0.3 is used.

Distribution and retention

(122) *Adults.* There is a fairly extensive literature on the distribution and retention of ^{223}Ra, ^{224}Ra, or ^{226}Ra in adult humans (Schlundt *et al.*, 1933; Norris *et al.*, 1955; Mays *et al.*, 1962, 1963; Schales, 1964; Miller and Finkel, 1965; Harrison *et al.*, 1967; ICRP, 1973; Schlenker *et al.*, 1982; Stather, 1990). These data, supplemented with information on the behaviour of radium in laboratory animals and on the physiologically similar element barium in humans, are described in Appendix A and summarized briefly here. Radium absorbed to blood from the gastrointestinal tract or lungs follows the general behaviour of calcium, although the rate of movement between plasma, bone and soft tissues, as well as the rates of urinary and faecal excretion, differ substantially for calcium and radium. A considerable fraction of radium leaving plasma passes into the intestines and is subsequently excreted in faeces. Secretion into the gastrointestinal tract is much greater in humans than in the commonly studied laboratory animals, leading to less

recycling of radium and substantially smaller net retention in humans than is generally observed in laboratory animals. There is substantial deposition and retention in bone and some in soft tissues. Much of the activity deposited in bone is returned to plasma within a few days or weeks, but a fraction is retained and is only slowly returned to plasma by bone remodelling and possibly other processes. In the fully mature human, skeletal retention may decrease from one-quarter or more of injected activity in the first day or two after injection to less than 10% at 1 month. Thereafter, skeletal retention gradually decreases to a level of about 0.5–1% by 25 years. Limited data for humans indicate that soft tissue radium may represent 20% or more of total body radium during the first several weeks after exposure but probably represents a much smaller percentage at times remote from exposure.

(123) The model developed by the Task Group on Alkaline Earth Metabolism in Adult Man (ICRP, 1973) is based to some extent on physiological considerations, but is also based in part on empirical fits to data, which do not allow incorporation of some information related to the age-specific biokinetics of radium. The model adopted here is reasonably consistent with the model of *ICRP Publication 20* (ICRP, 1973) but depicts more explicitly the anatomical compartments and physiological processes needed to model the age-specific uptake, retention and excretion of radium isotopes and their radioactive daughters in man. This model (Leggett, 1992) is described in Appendix A.

(124) *Children.* A limited amount of data on the biokinetics of radium in immature humans are available (ICRP, 1973; Parks *et al.*, 1978; Parks and Keane, 1983; Muth and Glöbel, 1983; Keane and Schlenker, 1987). These data must be supplemented with age-specific data for laboratory animals in order to construct a predictive model of the age-specific biokinetics of radium. Among laboratory animals used to study the behaviour of radium, the beagle has been studied most extensively (Lloyd *et al.*, 1976a,b, 1982, 1983a–d; Bruenger *et al.*, 1983; Bruenger and Lloyd, 1989).

(125) The combined information for humans and beagles indicates that retention of radium is greater in growing than mature animals and that much of the variation with age is due to elevated uptake and retention of radium by the immature skeleton. At times remote from exposure, skeletal burdens acquired during periods of growth tend to remain higher than those acquired by mature skeletons. Both deposition and removal of radium appear to be greater in areas of bone undergoing rapid remodelling than in areas of relatively slow remodelling. Greater deposition of radium in the younger skeleton results in less systemic radium available for excretion and distribution to soft tissues. The choice of age-specific biokinetic parameters for the model adopted here for radium is described in Appendix A.

Behaviour of decay products

(126) Except for isotopes of radon, thorium and actinium, radioactive progeny of radium radioisotopes produced in bone volume are assumed to follow the behaviour of the parent radionuclide until removed from bone volume. Outside bone volume, retention and translocation of decay products are described by a separate, characteristic biokinetic model for each element rather than the biokinetic model for the absorbed radium radioisotope. The age-dependent characteristic biokinetic model for lead is described in Appendix A. Age-independent models for polonium and bismuth as decay products are described in Chapter 11.

(127) Isotopes of thallium in the chains of interest here have half-lives of 4.2 min or less and relatively small branching fractions, and are of little importance with regard to

dosimetric estimates for intake of the radium parent. It is assumed here that thallium isotopes decay at their point of origin.

(128) Radon produced in soft tissues or bone surfaces is assumed to be removed to blood at the rate of 100 day^{-1} based on data of Harley *et al.* (1958), Van Dilla and Stover (1958) and Mays *et al.* (1958). Radon is assumed to migrate from shallow and deep bone volume to blood at rates of 1.5 and 0.36 day^{-1}, respectively. Resulting model predictions of the time-dependent increase in the ^{222}Rn:^{226}Ra ratio in bone are generally consistent with measurements in beagles and other laboratory animals during the first 2–3 months after injection with ^{226}Ra (Mays *et al.*, 1975; Lloyd and Bruenger, 1991). The model predicts a ^{222}Rn:^{226}Ra ratio in bone of about one-third for times remote from injection, which is consistent with data for human subjects (Mays *et al.*, 1975; ICRP, 1979). Radon entering blood from all compartments is assumed to be rapidly removed from the body by exhalation; a removal rate of 1 min^{-1} has been arbitrarily assigned.

(129) In beagles, ^{222}Rn escaped to a greater extent from immature bone than from mature bone during the first several months after injection of ^{226}Ra, but the radon/radium ratio in bone was independent of age at times remote from administration (Lloyd and Bruenger, 1991). In this report, the presumably conservative assumption is made that the biokinetics of radon produced in the body is independent of age.

(130) Thorium deposits mainly on bone surfaces, where it is tenaciously retained until buried in bone volume or removed to plasma and/or bone marrow by restructuring processes (Jee, 1964). In the liver it is deposited to a much smaller extent than americium and plutonium (Stover *et al.*, 1960; Singh *et al.*, 1983). In beagles, the rate of loss from the body over a period of years is roughly the same for thorium, plutonium and americium (Stover *et al.*, 1960; Lloyd *et al.*, 1984). Based on these considerations, a biokinetic model applied earlier to americium (ICRP, 1989) has been used as a starting point for modelling the biokinetics of thorium. However, modifications and simplifications of the americium model have been made to allow for known differences in distribution and excretion of the two elements, as well as for the considerable uncertainties regarding the biological behaviour of thorium as a daughter product. It is assumed that 7% of thorium leaving plasma goes to urine, 0.5% goes to the contents of the gastrointestinal tract and subsequently to faeces, 70% is evenly divided between cortical and trabecular bone surfaces, 4% deposits in the liver, 3% deposits in the kidneys, and the remainder goes to other tissues. Since children generally have a higher skeletal deposition of bone-seeking radioelements than do adults, it is assumed that 85% of thorium reaching plasma of persons up to 15 years of age is evenly divided between cortical and trabecular surfaces, and excretion and deposition in other organs are reduced by one-half to accommodate this higher skeletal deposition. In persons of all ages, thorium is assumed to move from the kidneys with a half-time of 2 years and from liver or other soft tissues to plasma with a half-time of 5 years. Thorium deposited on bone surfaces is assumed to be removed to plasma at the rate of bone turnover. Thorium born in shallow bone volume is assumed to move directly to deep bone volume and to be removed at the rate of bone turnover.

(131) It appears from the limited available data on biokinetics of actinium that this element deposits mainly in the skeleton and liver, is tenaciously retained in the skeleton, and is only slowly removed from the human body (Campbell *et al.*, 1956; Newton *et al.*, 1968; Taylor, 1970; Newton and Brown, 1974). The distribution of actinium in the skeleton of the rat is similar to that of americium and curium (Hamilton, 1948). In a

human subject internally contaminated with ^{227}Ac via a puncture wound it was estimated that 90% of actinium translocated from the wound was retained indefinitely (Newton *et al.*, 1968; Newton and Brown, 1974). At 3 years after the accident, activity appeared to be deposited primarily in bone with some involvement of liver, but after 9 years the contamination appeared to be almost exclusively skeletal. These findings suggest that the biokinetics of actinium is reasonably similar to that of thorium. Thus, the parameter values used in this report for actinium are the same as those applied to thorium. Since the only isotope of actinium appearing in the chains of interest here is ^{228}Ac, which has a half-life of only 6.1 h, application of the thorium model to actinium is virtually equivalent to assuming that this actinium isotope decays entirely at its site of origin.

Dose coefficients

(132) Dose coefficients derived from the biokinetic data summarized in Table A-1 are given in Tables 13.-1 to 13.-3.

Table 13.-1

Ingestion Dose Coefficients (Equivalent Dose (Sv/Bq) to Age 70 Years) for Ra-224 (T1/2 = 3.66 d)*

Age at intake	3 Months	1 Year	5 Years	10 Years	15 Years	Adult
Adrenals	2.7E-07	1.0E-07	5.1E-08	2.9E-08	1.3E-08	8.1E-09
Bladder Wall	2.7E-07	1.0E-07	5.2E-08	2.9E-08	1.4E-08	8.4E-09
Bone Surfaces	1.1E-04	2.3E-05	1.4E-05	1.1E-05	9.1E-06	1.8E-06
Brain	2.7E-07	1.0E-07	5.1E-08	2.9E-08	1.3E-08	8.0E-09
Breast	2.7E-07	1.0E-07	5.1E-08	2.9E-08	1.3E-08	8.0E-09
GI-Tract						
St Wall	3.2E-07	1.3E-07	6.6E-08	3.7E-08	1.9E-08	1.3E-08
SI Wall	3.2E-07	1.7E-07	8.3E-08	4.8E-08	2.4E-08	1.8E-08
ULI Wall	7.8E-07	6.1E-07	3.1E-07	1.8E-07	9.0E-08	8.2E-08
LLI Wall	1.7E-06	1.5E-06	7.7E-07	4.4E-07	2.3E-07	2.2E-07
Kidneys	1.3E-06	3.8E-07	2.0E-07	1.4E-07	1.1E-07	3.6E-08
Liver	9.1E-07	4.4E-07	2.3E-07	1.3E-07	6.6E-08	5.2E-08
Lungs	2.7E-07	1.0E-07	5.1E-08	2.9E-08	1.3E-08	8.0E-09
Muscle	2.7E-07	1.0E-07	5.1E-08	2.9E-08	1.3E-08	8.1E-09
Ovaries	2.7E-07	1.1E-07	5.3E-08	3.0E-08	1.4E-08	8.8E-09
Pancreas	2.7E-07	1.0E-07	5.1E-08	2.9E-08	1.3E-08	8.1E-09
Red Marrow	1.1E-05	1.7E-06	7.9E-07	6.8E-07	6.1E-07	1.7E-07
Skin	2.7E-07	1.0E-07	5.1E-08	2.9E-08	1.3E-08	8.1E-09
Spleen	1.9E-07	6.9E-08	3.4E-08	2.0E-08	1.1E-08	6.3E-09
Testes	2.7E-07	1.0E-07	5.1E-08	2.9E-08	1.3E-08	8.1E-09
Thymus	2.7E-07	1.0E-07	5.1E-08	2.9E-08	1.3E-08	8.0E-09
Thyroid	2.7E-07	1.0E-07	5.1E-08	2.9E-08	1.3E-08	8.0E-09
Uterus	2.7E-07	1.0E-07	5.2E-08	2.9E-08	1.3E-08	8.4E-09
Remainder	2.9E-07	1.1E-07	5.4E-08	3.1E-08	1.4E-08	8.6E-09
Effective Dose	2.8E-06	6.6E-07	3.5E-07	2.6E-07	2.0E-07	6.3E-08

GI-Tract Gastrointestinal Tract
St Stomach
SI Small Intestine
ULI Upper Large Intestine
LLI Lower Large Intestine

* In the biokinetic model for Ra parameter values for the adult apply to ages > 25 y. For radioisotopes of this element the dose coefficients for the adult are based on the 50-y integrated doses following an acute intake at age 25 y.

Table 13.-2

Ingestion Dose Coefficients (Equivalent Dose (Sv/Bq) to Age 70 Years) for Ra-226 (T1/2 = 1600 y)*

Age at intake	3 Months	1 Year	5 Years	10 Years	15 Years	Adult
Adrenals	5.3E-07	2.1E-07	1.3E-07	9.0E-08	8.1E-08	4.1E-08
Bladder Wall	5.3E-07	2.1E-07	1.3E-07	8.8E-08	7.3E-08	4.0E-08
Bone Surfaces	1.6E-04	2.9E-05	2.3E-05	4.0E-05	9.6E-05	1.3E-05
Brain	5.3E-07	2.1E-07	1.3E-07	9.0E-08	8.1E-08	4.1E-08
Breast	5.2E-07	2.1E-07	1.3E-07	8.7E-08	7.3E-08	4.0E-08
GI-Tract						
St Wall	5.4E-07	2.2E-07	1.3E-07	9.0E-08	7.4E-08	4.1E-08
SI Wall	5.4E-07	2.3E-07	1.4E-07	9.4E-08	7.8E-08	4.3E-08
ULI Wall	8.0E-07	4.1E-07	2.3E-07	1.5E-07	1.1E-07	6.7E-08
LLI Wall	1.7E-06	1.0E-06	5.4E-07	3.3E-07	2.1E-07	1.5E-07
Kidneys	6.9E-07	2.5E-07	1.6E-07	1.6E-07	2.4E-07	5.9E-08
Liver	3.8E-06	1.5E-06	7.8E-07	5.3E-07	4.0E-07	1.8E-07
Lungs	5.3E-07	2.1E-07	1.3E-07	8.8E-08	7.6E-08	4.0E-08
Muscle	5.3E-07	2.1E-07	1.3E-07	8.9E-08	7.7E-08	4.0E-08
Ovaries	5.3E-07	2.1E-07	1.3E-07	8.9E-08	7.7E-08	4.1E-08
Pancreas	5.3E-07	2.1E-07	1.3E-07	8.8E-08	7.6E-08	4.0E-08
Red Marrow	2.0E-05	3.0E-06	1.8E-06	2.5E-06	4.1E-06	8.9E-07
Skin	5.3E-07	2.1E-07	1.3E-07	8.8E-08	7.5E-08	4.0E-08
Spleen	6.7E-07	2.4E-07	1.6E-07	1.4E-07	1.9E-07	5.3E-08
Testes	5.2E-07	2.1E-07	1.3E-07	8.7E-08	7.3E-08	4.0E-08
Thymus	5.2E-07	2.1E-07	1.3E-07	8.8E-08	7.4E-08	4.0E-08
Thyroid	5.2E-07	2.1E-07	1.3E-07	8.8E-08	7.6E-08	4.0E-08
Uterus	5.3E-07	2.1E-07	1.3E-07	8.8E-08	7.4E-08	4.0E-08
Remainder	5.3E-07	2.1E-07	1.3E-07	9.0E-08	7.9E-08	4.1E-08
Effective Dose	4.7E-06	9.7E-07	6.2E-07	8.1E-07	1.5E-06	2.8E-07

GI-Tract Gastrointestinal Tract
St Stomach
SI Small Intestine
ULI Upper Large Intestine
LLI Lower Large Intestine

* In the biokinetic model for Ra parameter values for the adult apply to ages > 25 y. For radioisotopes of this element the dose coefficients for the adult are based on the 50-y integrated doses following an acute intake at age 25 y.

Table 13.-3

Ingestion Dose Coefficients (Equivalent Dose (Sv/Bq) to Age 70 Years) for Ra-228 (T1/2 = 5.75 y)*

Age at intake	3 Months	1 Year	5 Years	10 Years	15 Years	Adult
Adrenals	4.0E-06	9.7E-07	5.4E-07	4.0E-07	4.2E-07	1.1E-07
Bladder Wall	4.0E-06	9.6E-07	5.4E-07	4.0E-07	4.1E-07	1.1E-07
Bone Surfaces	1.0E-03	2.0E-04	1.4E-04	1.8E-04	2.5E-04	2.5E-05
Brain	3.9E-06	9.6E-07	5.4E-07	4.0E-07	4.2E-07	1.1E-07
Breast	3.9E-06	9.6E-07	5.4E-07	3.9E-07	4.0E-07	1.0E-07
GI-Tract						
St Wall	3.9E-06	9.6E-07	5.4E-07	3.9E-07	4.1E-07	1.0E-07
SI Wall	4.0E-06	9.7E-07	5.4E-07	4.0E-07	4.1E-07	1.1E-07
ULI Wall	5.0E-06	1.2E-06	6.6E-07	4.9E-07	5.5E-07	1.3E-07
LLI Wall	7.1E-06	1.7E-06	9.0E-07	6.9E-07	8.4E-07	1.8E-07
Kidneys	5.4E-05	9.3E-06	4.9E-06	5.4E-06	7.2E-06	7.5E-07
Liver	6.1E-05	1.3E-05	7.0E-06	6.0E-06	6.8E-06	1.3E-06
Lungs	3.9E-06	9.6E-07	5.4E-07	3.9E-07	4.1E-07	1.1E-07
Muscle	3.9E-06	9.6E-07	5.4E-07	3.9E-07	4.1E-07	1.1E-07
Ovaries	4.0E-06	9.7E-07	5.4E-07	4.0E-07	4.1E-07	1.1E-07
Pancreas	3.9E-06	9.6E-07	5.4E-07	3.9E-07	4.1E-07	1.1E-07
Red Marrow	1.2E-04	2.1E-05	1.2E-05	1.4E-05	1.7E-05	2.1E-06
Skin	3.9E-06	9.6E-07	5.4E-07	3.9E-07	4.1E-07	1.0E-07
Spleen	2.9E-06	6.8E-07	3.8E-07	3.1E-07	3.4E-07	8.5E-08
Testes	3.9E-06	9.6E-07	5.4E-07	3.9E-07	4.0E-07	1.0E-07
Thymus	3.9E-06	9.6E-07	5.4E-07	3.9E-07	4.1E-07	1.0E-07
Thyroid	3.9E-06	9.6E-07	5.4E-07	3.9E-07	4.1E-07	1.1E-07
Uterus	3.9E-06	9.6E-07	5.4E-07	3.9E-07	4.1E-07	1.1E-07
Remainder	4.8E-06	1.1E-06	6.0E-07	4.5E-07	4.8E-07	1.1E-07
Effective Dose	3.1E-05	6.0E-06	3.7E-06	4.1E-06	5.3E-06	6.6E-07

GI-Tract Gastrointestinal Tract
St Stomach
SI Small Intestine
ULI Upper Large Intestine
LLI Lower Large Intestine

* In the biokinetic model for Ra parameter values for the adult apply to ages > 25 y. For radioisotopes of this element the dose coefficients for the adult are based on the 50-y integrated doses following an acute intake at age 25 y.

References

Bruenger, F. W. and Lloyd, R. D. (1989) The influence of age at time of exposure to Ra-226 or Pu-239 on distribution, retention, postinjection survival and bone tumor induction in beagle dogs. In: *Thirty-fourth Annual Meeting of the Health Physics Society, Abstracts of Papers Presented at the Meeting*, June 25–29, 1989, Albuquerque Convention Center, Albuquerque, NM, Vol. 56 (Suppl. 1), 27. Pergamon Press, New York.

Bruenger, F. W., Smith, J. M., Atherton, D. R., Jee, W. S. S., Lloyd, R. D. and Stevens, W. (1983) Skeletal retention and distribution of Ra-226 and Pu-239 in beagles injected at ages ranging from 2 days to 5 years. *Health Phys.* **44**, 513–527.

Campbell, J. E., Robajdek, E. S. and Anthony, D. S. (1956) The metabolism of ^{227}Ac and its daughters ^{227}Th and ^{223}Ra by rats. *Radiat. Res.* **4**, 294–302.

Della Rosa, R. J., Goldman, M. and Wolf, H. G. (1967) *Uptake and Retention of Orally Administered Ra-226 and Ba-133. A Preliminary Report.* University of California at Davis, **472–114**, 40–41.

Hamilton, J. G. (1948) The metabolic properties of the fission products and actinide elements. *Rev. Mod. Phys.* **20**, 718–728.

Harley, J. H., Jelter, E. and Nelson, N. (1958) *Elimination of Radon from the Body.* USAEC Report HASL-32.

Harrison, G. E., Carr, T. E. F. and Sutton, A. (1967) Distribution of radioactive calcium, strontium, barium and radium following intravenous injection into a healthy man. *Int. J. Radiat. Biol.* **13**, 235–247.

ICRP (1973) *Alkaline Earth Metabolism in Adult Man,* ICRP Publication 20. Pergamon Press, Oxford.

ICRP (1979) *Limits for Intakes of Radionuclides by Workers: Part 1,* ICRP Publication 30. Pergamon Press, Oxford.

ICRP (1980) *Limits for Intakes of Radionuclides by Workers: Part 2,* ICRP Publication 30. Pergamon Press, Oxford.

ICRP (1989) *Age-dependent Doses to Members of the Public from Intake of Radionuclides: Part 1,* ICRP Publication 56. Pergamon Press, Oxford.

Jee, W. S. S. (1964) A critical survey of the analysis of microscopic distribution of some bone seeking radionuclides and assessment of absorbed dose. In: *Assessment of Radioactivity in Man,* Vol. II, pp. 369–393. IAEA, Vienna.

Keane, A. T. and Schlenker, R. A. (1987) Long-term loss of radium in 63 subjects exposed at ages 6 to 46. In: *Age-related Factors in Radionuclide Metabolism and Dosimetry* (G. B. Gerber, H. Métivier and H. Smith, Eds), pp. 127–135. Martinus Nijhoff Publishers, Dordrecht.

Leggett, R. W. (1992) A generic biokinetic age-specific model for calcium-like elements. Paper presented at the Workshop on "Age-dependent Factors in the Biokinetics and Dosimetry of Radionuclides", Schloß Elmau, Germany, 5–8 November 1991. *Radiat. Prot. Dosim.* **41**, 183–198.

Lloyd, R. D. and Bruenger, F. W. (1991) Rn:Ra ratios in bone of beagles injected with ^{226}Ra. *Health Phys.* **60**, 567–568.

Lloyd, R. D., Mays, C. W. and Atherton, D. R. (1976a) Distribution of injected Ra-226 and Sr-90 in the beagle skeleton. *Health Phys.* **30**, 183–189.

Lloyd, R. D., Mays, C. W., Atherton, D. R., Taylor, G. N. and Van Dilla, M. A. (1976b) Retention and skeletal dosimetry of injected Ra-226, Ra-228, and Sr-90 in beagles. *Radiat. Res.* **66**, 274–287.

Lloyd, R. D., Mays, C. W., Taylor, G. N., Atherton, D. R., Bruenger, F. W. and Jones, C. W. (1982) Radium-224 retention, distribution, and dosimetry in beagles. *Radiat. Res.* **92**, 280–295.

Lloyd, R. D., Bruenger, F. W., Jones, C. W., Taylor, G. N. and Mays, C. W. (1983a) Radium retention in mature beagles injected at 5 years of age. *Radiat. Res.* **94**, 210–216.

Lloyd, R. D., Bruenger, F. W., Mays, C. W. and Jones, C. W. (1983b) Skeletal radon-to-radium ratios in neonatal, juvenile and mature beagles and in adult St. Bernards. *Health Phys.* **44**, 61–63.

Lloyd, R. D., Jones, C. W., Bruenger, F. W., Atherton, D. R. and Mays, C. W. (1983c) Radium retention and dosimetry in juvenile beagles. *Radiat. Res.* **94**, 295–304.

Lloyd, R. D., Taylor, G. N., Jones, C. W. and Mays, C. W. (1983d) Radium retention and dosimetry in the St. Bernard. *Radiat. Res.* **95**, 150–157.

Lloyd, R. D., Mays, C. W., Jones, C. W., Atherton, D. R., Bruenger, F. W., Shabestari, L. R. and Wrenn, M. E. (1984) Retention and dosimetry of injected Am-241 beagles. *Radiat. Res.* **110**, 564–575.

Maletskos, C. J., Keane, A. T., Trelles, N. C. and Evans, R. D. (1966) The metabolism of intravenously administered radium and thorium in human beings and the relative absorption from the human gastrointestinal tract of radium and thorium in simulated radium dial paints. Annual Report MIT-952-3, Radioactivity Center, Institute of Technology, Cambridge, MA, pp. 202–317.

Maletskos, C. J., Keane, A. T., Trelles, N. C. and Evans, R. D. (1969) Retention and absorption of Ra-224 and Th-234 and some dosimetric considerations of Ra-224 in human beings. In: *Delayed Effects of Bone-seeking Radionuclides* (C. W. Mays, W. S. S. Jee, R. D. Lloyd, B. J. Stover, J. H. Dougherty and G. N. Taylor, Eds), pp. 29–49. University of Utah Press, Salt Lake City, UT.

Mays, C. W., Haldin, R. and Van Dilla, M. A. (1958) Thoron exhalation in radiothorium-burdened beagles. *Radiat. Res.* **9**, 438–444.

Mays, C. W., Christensen, W. R., Lloyd, R. D., Atherton, D. R., Parmley, W. W. and Pitchford, G. S. (1962) Radium studies in a man soon after injection. In: *Research in Radiobiology.* University of Utah, **COO-226**, 71–77.

Mays, C. W., Lloyd, R. D., Christensen, W. R., Atherton, D. R. and Pitchford, G. S. (1963) Radium metabolism in a man. *Radiat. Res.* **19**, 210 (abstract).

Mays, C. W., Lloyd, R. D. and Van Dilla, M. A. (1975) Fractional radon retention in bone. *Health Phys.* **29**, 761–765.

Miller, C. E. and Finkel, A. J. (1965) A reexamination of retention patterns in patients who received radium by multiple injections 33 years earlier. In: *Health Division Gamma-ray Spectroscopy Group Annual Report, July 1964 through June 1965.* Argonne National Laboratory, **ANL-7217**, 5–90.

Muth, H. and Glöbel, B. (1983) Age dependent concentration of Ra-226 in human bone and some transfer factors from diet to human tissues. *Health Phys.* **44**, 113–121.

NEA/OECD (1988) Committee on Radiation Protection and Public Health. *Expert Group on Gut Transfer Factors.* NEA/OECD Report, Paris.

Newton, D. and Brown, D. A. (1974) The long-term retention of systemic protactinium-231 and actinium-227. *Health Phys.* **27**, 459–467.

Newton, D., Rundo, J. and Sandalls, F. J. (1968) A case of internal contamination with ^{231}Pa and ^{227}Ac via a puncture wound. In: *Diagnosis and Treatment of Deposited Radionuclides* (H. A. Kornberg and W. D. Norwood, Eds), pp. 521–533. Excerpta Medica Foundation, Amsterdam.

Norris, W. P., Speckman, T. W. and Gustafson, P. F. (1955) Studies on metabolism of radium in man. *Am. J. Roentgenol., Radium Ther. Nucl. Med.* **73**, 785–802.

Parks, N. J. and Keane, A. T. (1983) Consideration of age-dependent radium retention in people on the basis of the beagle model. *Health Phys.* **44**, 103–112.

Parks, N. J., Pool, R. R., Williams, J. R. and Wolf, H. G. (1978) Age and dosage-level dependence of radium retention in beagles. *Radiat. Res.* **75**, 617–632.

Schales, F. (1964) The excretion of thorium X and its daughter products after intravenous injection in man. In: *Assessment of Radioactivity in Man*, Vol. II, pp. 267–276. IAEA, Vienna.

Schlenker, R. A., Keane, A. T. and Holtzman, R. B. (1982) The retention of Ra-226 in human soft tissue and bone; implications for the ICRP 20 alkaline earth model. *Health Phys.* **42**, 671–693.

Schlundt, H., Nerancy, J. T. and Morris, J. P. (1933) The detection and estimation of radium in living persons. IV. The retention of soluble radium salts administered intravenously. *Am. J. Roentgenol., Radium Ther.* **30**, 515–522.

Seil, H. A., Viol, C. H. and Gordon, M. A. (1915) The elimination of soluble radium salts taken intravenously and per os. *Radium* **5**, 40–44 (also see ICRP Publication 23, *Report of the Task Group on Reference Man*, 1975, p. 405).

Singh, N. P., Wrenn, M. E. and Ibrahim, S. A. (1983) Plutonium concentration in human tissues: comparison to thorium. *Health Phys.* **44**, Suppl. 1, 469–476.

Stather, J. W. (1990) The behaviour, effects and radiation dosimetry of radium in man. In: *IAEA Technical Reports Series*, No. 310, Vol. 2, pp. 297–343. IAEA, Vienna.

Stehney, A. F. and Lucas, H. F. (1956) Studies on the radium content of humans arising from the natural radium of their environment. *Peaceful Uses of Atomic Energy (Proc. Int. Conf. Geneva, 1955)*, Vol. 11, p. 49. United Nations, New York.

Stover, B. J., Atherton, D. R., Keller, N. and Buster, D. S. (1960) Metabolism of the Th-228 decay series in adult beagle dogs. *Radiat. Res.* **12**, 657–671.

Taylor, D. M. (1970) The metabolism of actinium in the rat. *Health Phys.* **19**, 411–418.

Taylor, D. M., Bligh, P. H. and Duggan, M. H. (1962) The absorption of calcium, strontium, barium, and radium from the gastrointestinal tract of the rat. *Biochem. J.* **83**, 25–29.

Van Dilla, M. A. and Stover, B. J. (1956) On the role of radiothorium (Th-228) in radium poisoning. *Radiology* **66**, 400–441.

APPENDIX A: AGE-SPECIFIC BIOKINETIC MODELS FOR THE ALKALINE EARTH ELEMENTS AND LEAD

Age-specific Biokinetic Models for Alkaline Earth Elements

Comparative behaviours of the alkaline earth elements

(A1) The alkaline earth elements strontium, barium and radium follow the movement of calcium in the body but exhibit different patterns of distribution from that of calcium due to discrimination by biological membranes and bone mineral. In general, strontium is a better quantitative tracer for calcium than are the heavier elements, barium and radium (Harrison *et al.*, 1967). Experimental data indicate, however, that strontium is less effectively absorbed from the intestines, more effectively excreted by the kidneys, and perhaps less readily incorporated into new bone than calcium (Bauer *et al.*, 1955; Comar and Wasserman, 1958, 1964; Spencer *et al.*, 1960; Barnes, *et al.*, 1961; Cohn *et al.*, 1963; Decker *et al.*, 1964; Harrison *et al.*, 1967).

(A2) Calcium and strontium are excreted primarily in urine, while barium and radium are excreted mainly in faeces, with the total excretion rate soon after injection being higher for the heavier alkaline earths than for calcium and strontium (Harrison *et al.*, 1967). All four elements have similar skeletal uptake and distribution at early times after injection (Ellsasser *et al.*, 1969; Wood *et al.*, 1970; Liniecki, 1971; Stather, 1974; Lloyd *et al.*, 1976a). Within a few months after administration of isotopes of calcium, strontium, barium or radium, nearly all of the remaining total-body activity is associated with bone mineral (ICRP, 1973; Schlenker *et al.*, 1982).

(A3) Radium and barium exhibit fairly similar kinetics in the body, particularly at early times after injection (Stather, 1990). In a study of the fate of ^{226}Ra and ^{133}Ba acutely ingested by eight beagles from 43 to 1500 days of age, Della Rosa *et al.* (1967) found that these two radionuclides were absorbed and retained with nearly the same efficiency in each animal, with 30-day retention of barium being slightly greater as an average than that of radium. Data for a healthy 60-year-old male human injected with ^{223}Ra and ^{133}Ba indicate similar retention of the two radionuclides in plasma and total body for several days after injection and a slightly more rapid decline of whole-body ^{223}Ra after a few weeks (Harrison *et al.*, 1967; Newton *et al.*, 1977). In human studies, radium and barium were excreted primarily in faeces (Schales, 1964; Harrison *et al.*, 1967; Maletskos *et al.*, 1969; Korsunskii *et al.*, 1981) and showed fairly similar faecal excretion rates for at least a month after injection (Harrison *et al.*, 1967). Barium may be eliminated in urine at a greater rate than radium (Harrison *et al.*, 1967), but urinary excretion constitutes only a small fraction of total excretion of both elements in humans (Harrison *et al.*, 1967; Korsunskii *et al.*, 1981; Newton *et al.*, 1991). Finally, collective experimental data on radium and barium in humans indicate grossly similar whole-body retention patterns for the two elements, the main apparent difference being that whole-body retention of radium falls slightly faster than that of barium in the early weeks or months after exposure (Schlundt *et al.*, 1933; Norris *et al.*, 1955; Mays *et al.*, 1962, 1963; Miller and Finkel, 1965; Le Roy *et al.*, 1966; Harrison *et al.*, 1967; Rundo, 1967; ICRP, 1973; Newton *et al.*, 1977; Erre *et al.*, 1980; Schlenker *et al.*, 1982; Newton *et al.*, 1991).

An overview of the models for the alkaline earth elements

(A4) These models are described in detail elsewhere (Leggett, 1992) and summarized here. The following general behaviour of strontium, barium and radium is depicted. Activity entering blood (plasma) from the lungs or gastrointestinal tract is retained by bone surfaces and soft tissues or excreted in the urine and faeces (Fig. A1). Much of the strontium, barium and radium deposited on bone surfaces is returned to plasma within a few days, but a fraction migrates to regions of bone that lose activity to plasma much more slowly. All activity leaving the soft tissue compartments is assumed to be returned to plasma. Activity returned to plasma is assumed to be redistributed among tissues and excreta according to the same parameter values as for the original input to plasma.

(A5) The present model differs in form from the alkaline earth model for adult man described in *ICRP Publication 20* (ICRP, 1973) but was strongly influenced by that model in several respects, particularly with regard to the description of uptake and removal of alkaline earth elements by bone. As in the earlier alkaline earth model, bone

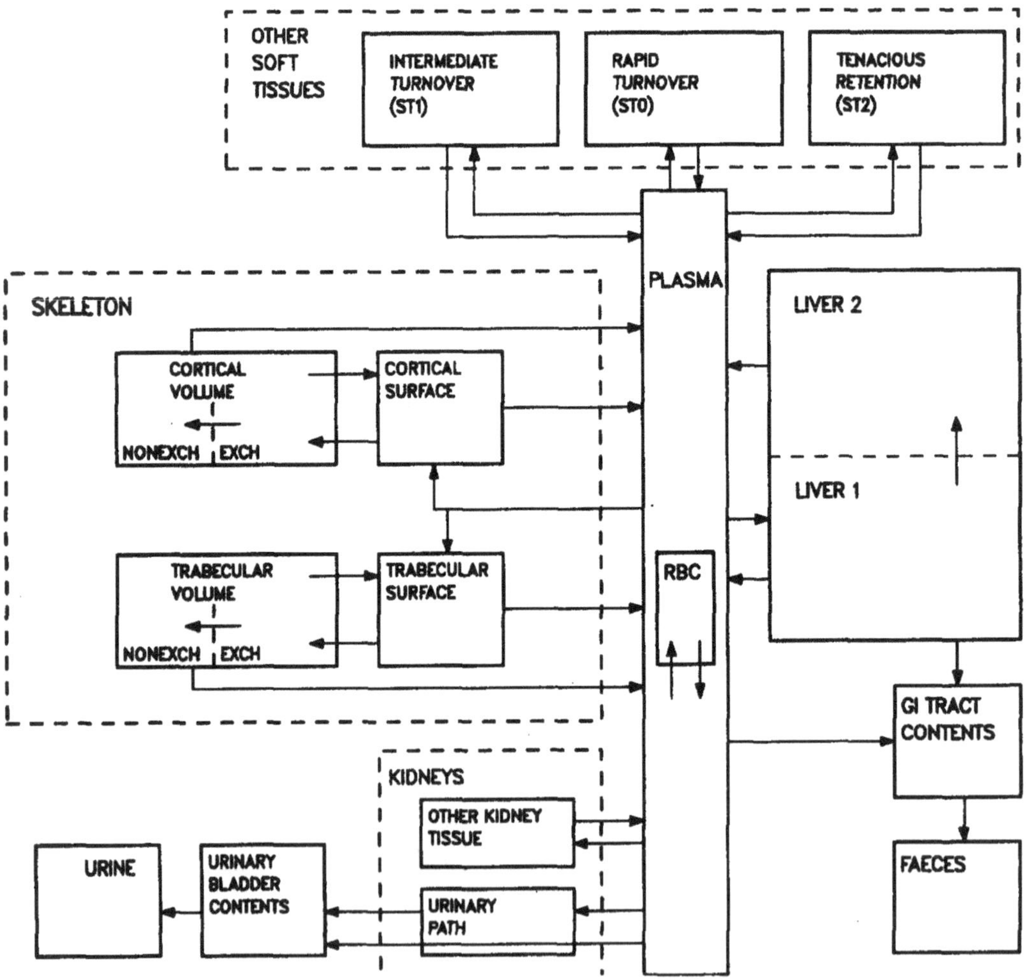

Fig. A1. Diagram of the biokinetic model for alkaline earth elements and lead.

is divided into cortical and trabecular components, each of which is further divided into bone surface and bone volume; rapidly exchangeable activity in bone is assumed to reside on bone surface, which is treated as a uniformly mixed compartment that exchanges activity with blood plasma; and biological removal from bone volume is assumed to occur by processes of diminution and resorption. In contrast to the earlier alkaline earth model, some parameter values in the present model vary with age; redeposition of biologically removed activity is treated explicitly; and diminution of activity in bone volume (represented as a power-function component in the earlier model) is assumed to be a first-order process.

(A6) In the present model, cortical and trabecular bone volume are each assumed to consist of two subcompartments, representing relatively exchangeable activity and relatively non-exchangeable activity in bone volume and referred to as exchangeable and non-exchangeable bone volume, respectively. These compartments are denoted by EXCH and NONEXCH, respectively, in Fig. A1.

(A7) Activity entering the skeleton is assumed to deposit initially on bone surfaces but to return to plasma or migrate to exchangeable bone volume within a few days. A portion of activity leaving exchangeable bone volume is assumed to return to bone surfaces and the rest is assigned to non-exchangeable bone volume from which it is gradually removed to plasma by bone resorption.

(A8) No attempt is made here to model all of the paths of movement known or thought to be involved in transfer of activity among plasma, bone surfaces, bone fluids and bone volume. The model is intended only to provide reasonably accurate predictions of the time-dependent activity in bone surfaces and bone volume after injection of isotopes of strontrium, barium or radium into plasma, using a minimal number of compartments and first-order transfers between compartments.

Selection of parameter values for alkaline earth elements

(A9) Because of the qualitative similarities in the biological behaviour of strontium, barium and radium, common compartments and paths of movement are used in the age-specific biokinetic models for these elements (Fig. A1). (The model structure shown in Fig. A1 includes some features not used for the alkaline earth elements but needed for consideration of lead and potentially other bone-volume-seeking elements.) Selection of parameter values for each element is based as far as practical on element-specific data but also relies to some extent on observed qualitative similarities or differences in the biological behaviours of the three elements. In particular, much use is made of the close similarities in the observed biological behaviours of barium and radium and of known differences between the lighter and heavier alkaline earth elements. The model for the alkaline earth elements was based on the parameter values originally developed for radium with subsequent modification for application to barium and strontium. Parameter values have been derived for the 100-day-old infant, the 1-, 5-, 10-, 15-year-old child and for the mature adult of age greater than 25 years. The values for the 100-day-old infant have been assigned to the 3-month-old infant, and the values for the mature adult have been assigned to the adult. The details of the models are described in the order of their development.

1. Radium

(A10) *Adults.* The biokinetic models of Leggett (1992) for radium, barium and strontium address potential differences with age during adulthood. Whole body retention of radium is generally higher for persons exposed in their early twenties than in later years (Schlundt *et al.*, 1933; Norris *et al.*, 1955; Miller and Finkel, 1965; ICRP, 1973; Parks *et al.*, 1978; Parks and Keane, 1983). In the simplified versions given here, parameter values for adults are only for a "mature adult", defined to be at least 25 years of age.

(A11) There have been many indirect measurements and some autopsy measurements of radium isotopes in briefly exposed humans (Schlundt *et al.*, 1933; Norris *et al.*, 1955; Mays *et al.*, 1963; Miller and Finkel, 1965; Harrison *et al.*, 1967; ICRP, 1973; Parks *et al.*, 1978; Harrison, 1981; Schlenker *et al.*, 1982; Parks and Keane, 1983; Keane and Schlenker, 1987). These data can be used to estimate the distribution and retention of radium in the body from a few days to several decades after acute exposure. For estimation of transfer rates controlling movement of radium in the body during the first few days after intake, it is useful to supplement human retention data on radium with barium injection data for humans (Korsunskii *et al.*, 1981) and radium or barium injection data for beagles (Cuddihy and Griffith, 1972; Lloyd *et al.*, 1976a,b, 1982, 1983a). In extrapolations of the beagle data to humans, consideration must be given to the substantially lower faecal excretion rate and much greater recycling of the heavy alkaline earths in beagles than in humans, as indicated, for example, by data of Van Dilla *et al.* (1958), Harrison *et al.* (1967), Cuddihy and Griffith (1972) and Newton *et al.* (1991).

(A12) In young adult beagles, soft tissues contained about 62% of the total-body content of injected ^{224}Ra at 1 h, 29% at 1 day and 12% at 7 days (Lloyd *et al.*, 1982). In a re-evaluation of data on ^{226}Ra in human soft tissues, Schlenker *et al.* (1982) estimated that soft tissue retention rises to about 58% of whole body retention at 18 days after single intake and then falls steadily to 33% at 100 days and 6% at 1000 days. The estimates of Schlenker and coworkers represent an interpretation of available human data within certain constraints of the ICRP alkaline earth model (ICRP, 1973), and seem likely to overestimate soft tissue retention at early times (Stather, 1990); a more direct fitting procedure would yield somewhat lower estimates at early times.

(A13) It has been estimated from autopsy measurements of natural ^{226}Ra concentrations in adult humans that soft tissues contain 10–30% of total-body ^{226}Ra (Hursh and Lovaas, 1963; Rajewsky *et al.*, 1965; Maletskos *et al.*, 1969; ICRP, 1973; Qiyue *et al.*, 1988). These estimates were based on mean values or pooled samples for several subjects. Schlenker *et al.* (1982) pointed out that these two methods of estimation may be inappropriate since measured ^{226} Ra concentrations are likely to be asymmetrically distributed. Using median values of ^{226}Ra to Ca ratios obtained from the literature, they estimated that soft tissues contain 5.5–6% of the natural ^{226}Ra in the total body.

(A14) In the present model for adults, selection of parameter values for the soft tissue compartments (Table A-1) were set to yield reasonable agreement with: (1) ^{224}Ra and ^{226}Ra injection data for beagles for 1 h to 7 days after administration (Atherton *et al.*, 1965; Lloyd *et al.*, 1982); (2) measurements of ^{226}Ra at 5 to 426 days after single injections into three terminally ill human subjects and at 3 to 53 years after extended ingestion by 14 human subjects (see Schlenker *et al.*, 1982) (Figs A2 and A3); and (3) the assumption that about 5–6% of total-body radium is in all soft tissues after several decades of constant intake (Schlenker *et al.*, 1982). The kidneys are not addressed explicitly but are assumed to be part of "other soft tissues". The liver is assumed to be kinetically distinct from other soft tissues (Table A-1).

Table A-1 Age-specific transfer rates (d^{-1}) for radium model

	Age					
	3 mo	1 y	5 y	10 y	15 y	Adult
plasma to urinary bladder contents	0.202	0.444	0.488	0.355	0.210	0.606
plasma to ULI contents	7.26	16.0	17.43	12.78	7.55	21.79
plasma to trabecular bone surface	10.5	6.30	6.22	9.88	14.45	9.72
plasma to cortical bone surface	42.0	25.2	21.78	29.32	37.35	7.78
plasma to liver 1	0.117	0.257	0.280	0.205	0.121	0.350
plasma to ST0	7.56	16.63	18.14	13.31	7.86	22.68
plasma to ST1	2.33	5.13	5.60	4.11	2.43	7.00
plasma to ST2	0.0233	0.0513	0.0560	0.0411	0.0243	0.070
bone surface to plasma	0.578	0.578	0.578	0.578	0.578	0.578
bone surface exch[a] bone volume	0.116	0.116	0.116	0.116	0.116	0.116
exch bone volume to bone surface	0.0185	0.0185	0.0185	0.0185	0.0185	0.0185
exch bone volume to nonexch volume	0.0046	0.0046	0.0046	0.0046	0.0046	0.0046
nonexch[a] trabecular volume to plasma	0.00822	0.00288	0.00181	0.00132	0.000959	0.000493
nonexch cortical volume to plasma	0.00822	0.00288	0.00153	0.000904	0.000521	0.0000821
liver 1 to plasma	0.0139	0.0139	0.0139	0.0139	0.0139	0.0139
ST0 to plasma	2.52	5.54	6.05	4.44	2.62	7.56
ST1 to plasma	0.693	0.693	0.693	0.693	0.693	0.693
ST2 to plasma	0.00038	0.00038	0.00038	0.00038	0.00038	0.00038
f_1	0.6	0.3	0.3	0.3	0.3	0.2

[a]exch = exchangeable, nonexch = nonexchangeable

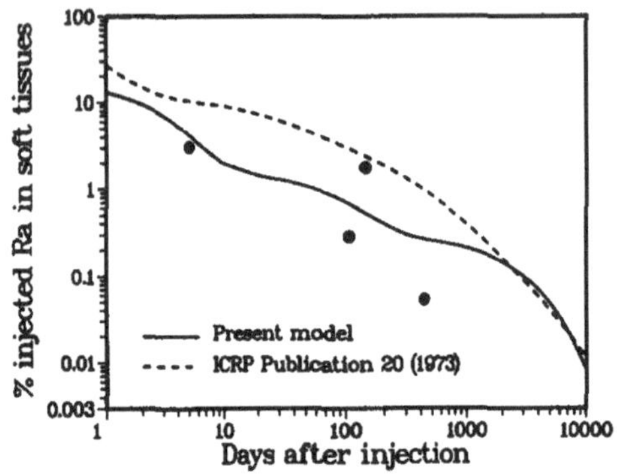

Fig. A2. The radium content of soft tissues of a mature adult as a function of time after injection, based on the present model, the alkaline earth model of *ICRP Publication 20* (ICRP, 1973), and data for terminally ill human subjects who received a single injection of ^{226}Ra (Schlenker *et al.*, 1982).

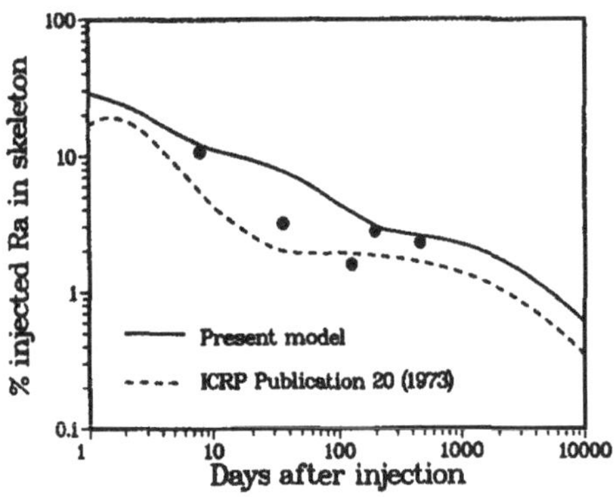

Fig. A3. The radium content of the skeleton of a mature adult as a function of time after injection, based on the present model, the alkaline earth model of *ICRP Publication 20* (ICRP, 1973), and data for terminally ill human subjects who received a single injection of ^{226}Ra (Schlenker *et al.*, 1982).

(A15) The model predicts that soft tissues contain less than 20% of total-body radium beyond 1 day after injection. Figure A2 shows the soft tissue content of radium as a function of time after entry into the systemic circulation based on the parameter values for adults given in Table A-1. These predictions for soft tissues are substantially lower than those of the previous ICRP model for radium (ICRP, 1973) for the first few years after injection.

(A16) The estimated net rate of exchange between blood and bone surfaces in adults (Table A-1) was based on injection data for ^{224}Ra and ^{226}Ra in beagles (Lloyd *et al.*, 1976a, b, 1982, 1983a) and comparative data on the kinetics of the alkaline earth elements in man (Leggett, 1992). Rates of movement from bone surfaces to exchangeable bone volume and from exchangeable bone volume to non-exchangeable bone volume or to bone surfaces were set to yield reasonable agreement with intermediate- and long-term retention data for radium in the human skeleton (Schlenker *et al.*, 1982) and whole body (Schlundt *et al.*, 1933; Norris *et al.*, 1955; Mays *et al.*, 1963; Miller and Finkel, 1965; Harrison *et al.*, 1967; ICRP, 1973; Parks *et al.*, 1978; Harrison, 1981; Parks and Keane, 1983; Keane and Schlenker, 1987). Figure A3 gives

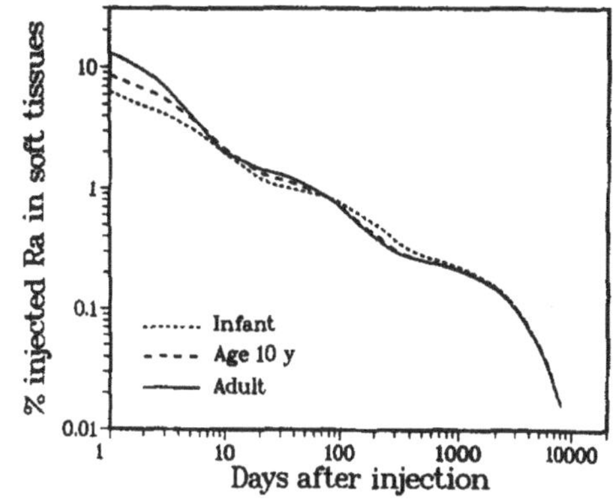

Fig. A4. Model predictions of the radium content of the soft tissues as a function of age at injection and time after injection.

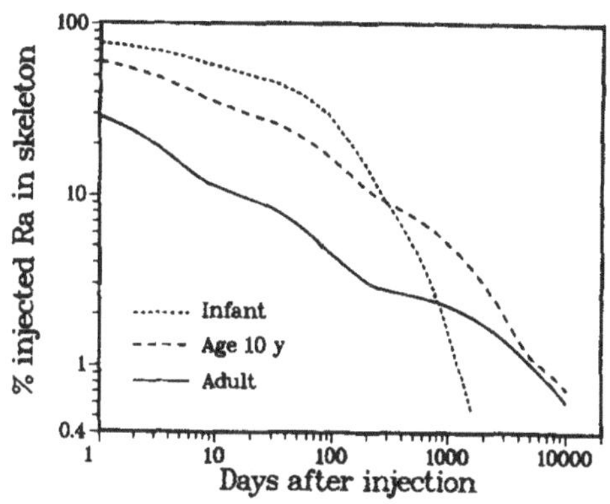

Fig. A5. Model predictions of the radium content of the skeleton as a function of age at injection and time after injection.

the skeleton content of radium as a function of time after entry into the systemic circulation based on the parameter values for adults given in Table A-1.

(A17) Systemic radium is assumed to be excreted in faeces after secretion from blood to the contents of the upper large intestine. The rate of movement from blood to the GI tract (Table A-1) was estimated from data of Harrison *et al.* (1967) and Maletskos *et al.* (1969) for adult humans. The rate of removal of radium in urine was based on an assumed faecal to urinary excretion ratio of 36:1 over an extended period (Schales, 1964; Harrison *et al.*, 1967; Maletskos *et al.*, 1969). The rate of removal of radium from plasma (Table A-1) was based on data on healthy human subjects injected with radioisotopes of radium and barium (Harrison *et al.*, 1967; Newton *et al.*, 1991) and the assumption that the removal rate from plasma is the same for radium and barium.

(A18) *Children.* There is evidence that retention of radium is greater in growing than mature animals and that much of the variation with age is due to elevated uptake of radium by the immature skeleton. In mice, uptake of injected ^{224}Ra by the skeleton decreased continuously with age at injection, from about 50% of the injected amount in young animals to about 25% in mature animals, as determined at 48 h after injection (Müller, 1987). In injection studies with newborn, juvenile (3 months), young adult (17–19 months) and old (5 years) beagles, skeletal retentions of ^{226}Ra at 1–2 weeks after injection were about 85%, 65%, 30–50%, and 30–50%, respectively, of the administered amount (Lloyd *et al.*, 1976a, b, 1982, 1983a–d; Bruenger *et al.*, 1983, 1989; Bruenger and Lloyd, 1989). At extended times, the skeletal burden remained higher in beagles injected as juveniles than in those injected as adults. Also, the skeletal burden remained noticeably greater in animals injected as young adults than in those injected at age 5 years. In beagles, deposition of radium apparently is greater, and its removal is more rapid, in areas of bone undergoing rapid remodelling than in areas of relatively slow remodelling (Jee and Polig, 1989).

(A19) Parks *et al.* (1978) and Parks and Keane (1983) concluded that initial retention of ^{226}Ra in beagles and humans varies in proportion to the calcium addition rate at the time of intake. Muth and Glöbel (1983) measured the concentration of ^{226}Ra in bone samples from subjects of all ages and determined two clear maxima in the plot of concentration vs age, one between 0 and 1 year and the other between 10 and 16 years. They noted that these are the ages when the human skeleton goes through its first and second rapid growth phases. Earlier measurements (Hallden *et al.*, 1963; Fisenne *et al.*, 1981) had not indicated any clear variation with age in the ^{226}Ra concentration in vertebral samples, but those data were less extensive and more scattered than the data of Muth and Glöbel.

(A20) In the model for radium (as well as those for barium and strontium), differences with age in parameter values are based on the following assumptions.

1. Fractional deposition on trabecular or cortical bone surfaces is proportional to the calcium addition rate (g Ca day^{-1}) for the given bone type. For age 1 year or greater, estimates of the calcium addition rates for trabecular and cortical bone are taken from a paper by Leggett *et al.* (1982). The calcium addition rate for 3-month-old infants has been increased from the value given in that paper, in view of the particularly large uncertainty in the estimated calcium addition rate for infants.
2. The rate of loss from non-exchangeable bone volume is equal to the age-specific rate of bone resorption, as given in *ICRP Publication 56* (ICRP, 1989).
3. The rates of loss from bone surfaces and exchangeable bone volume (i.e. the sum of transfer rates to all destinations) are independent of age. As described later, however, fractional transfer of strontium from bone surfaces to exchangeable bone volume is assumed to be smaller in children than in adults. This modification is not made for the heavier alkaline earths due to a paucity of information.
4. At all ages, deposition in soft tissues and excreta is proportional to the corresponding fractions for mature adults. For children, the amount left over after subtraction of bone deposition is divided into deposition fractions for soft tissues and excreta in proportion to the corresponding deposition fractions for mature adults. For example, if the deposition fraction for bone surfaces is 0.25 in mature adults and 0.75 in infants, then deposition fractions for soft tissue compartments and excretion pathways for infants are $(1.0-0.75)/(1.0-0.25)=1/3$ times the corresponding deposition fractions for adults. This approach is reasonably consistent with age-specific data on radium in beagles (Atherton *et al.*, 1965; Lloyd *et al.*, 1983) and barium in humans (Bauer *et al.*, 1957).

5. Removal rates from soft tissue compartments are the same in children as in adults, with the exception of soft tissue compartment ST0. The removal rate from this compartment is set so that ST0 contains three times as much activity as plasma at equilibrium at all ages (Leggett, 1992).

(A21) Data for beagles indicate faster removal from plasma in immature than in mature animals (Bruenger *et al.*, 1983). Because the assumed removal rate from blood in adults is already very high (70 day^{-1}), however, an assumption of faster removal from blood at younger ages would have virtually no effect on dosimetric estimates for important radium isotopes. Thus, the rate of removal of radium from blood is assumed to be independent of age.

(A22) Figures A4 and A5 give the soft tissue and skeleton content of radium for infants, 10-year-old children, and adults after entry into the systemic circulation based on the parameter values given in Table A-1.

2. Barium

(A23) *Adults.* The biological behaviour of injected or ingested barium has been investigated in several human subjects (Bauer *et al.*, 1957; LeRoy *et al.*, 1966; Harrison *et al.*, 1967; Korsunskii *et al.*, 1981; Newton *et al.*, 1991) and in a variety of laboratory animals (Richmond *et al.*, 1960, 1962a, b; Bligh and Taylor, 1963; Farnham and Rowland, 1965; Della Rosa *et al.*, 1967; Ellsasser *et al.*, 1969; Hardy *et al.*, 1969; Wood *et al.*, 1970; Cuddihy and Griffith, 1972; Stather, 1974; Domanski *et al.*, 1969, 1980). Some of these studies involve direct comparisons of the biological behaviour of barium and radium (Della Rosa *et al.*, 1967; Harrison *et al.*, 1967; Hardy *et al.*, 1969; Wood *et al.*, 1970). Based on the above studies and on the information for radium discussed earlier, the following differences in the behaviours of barium and radium in the adult human are assumed.

1. The rate of clearance from plasma to urine is nearly four times higher for barium than radium (Harrison *et al.*, 1967; Newton *et al.*, 1991). The rate of clearance of barium from plasma to the gastrointestinal tract was set so that the total rate of clearance from blood to excreta is the same for barium and radium (Harrison *et al.*, 1967).
2. The removal half-time of barium from exchangeable bone volume is 50 days, compared with 30 days for radium. In the adult, 30% of barium leaving exchangeable bone volume goes to non-exchangeable bone volume, compared with 20% of radium, giving a "discrimination factor" of 0.67 with regard to the relative propensities of skeletal radium and barium to become "permanently fixed" in bone (i.e. removable only by resorption). A similar discrimination between barium and radium is assumed for all ages. The assumed discrimination is reasonably consistent with data of Stark (1968) for relative binding of these elements by synthetic hydroxyapatite crystals, an *in vitro* analogue of bone mineral.
3. Available data do not indicate elevated uptake or retention of barium by the liver (Loutit and Russell, 1961; Bligh and Taylor, 1963; Wood *et al.*, 1970; Korsunskii *et al.*, 1981; Newton *et al.*, 1991). Thus this organ is not treated separately from other soft tissues. A larger fraction of barium than radium is assigned to the soft tissue compartment ST2 (representing long-term retention), however, to give reasonable agreement with the estimate of Schlenker *et al.* (1982) that almost 5% of the body's natural barium resides in soft tissues in the average adult.

(A24) *Children.* Age dependence in the biokinetics of barium has been investigated in laboratory animals (Farnham and Rowland, 1965; Ellsasser *et al.*, 1969; Hardy *et al.*, 1969; Cuddihy and Griffith, 1972; Stather, 1974; Domanski *et al.*, 1980) and in human infants, children and adults (Bauer *et al.*, 1957). Available data suggest that differences with age in the biokinetics of barium are entirely analogous with those discussed earlier for radium. Hence, the method of extension of the adult model for barium to children is the same as that described above for radium. Figures A6 and A7 give the soft tissue and skeleton content of barium for infants, 10-year-old children and adults after entry into the systemic circulation based on the parameter values given in Table A-2.

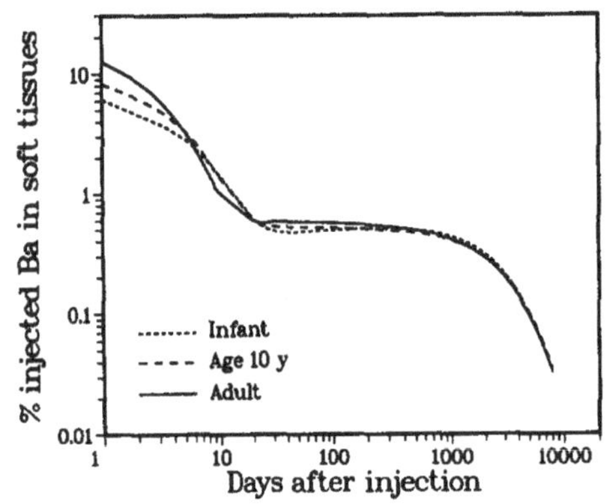

Fig. A6. Model predictions of the barium content of the soft tissues as a function of age at injection and time after injection.

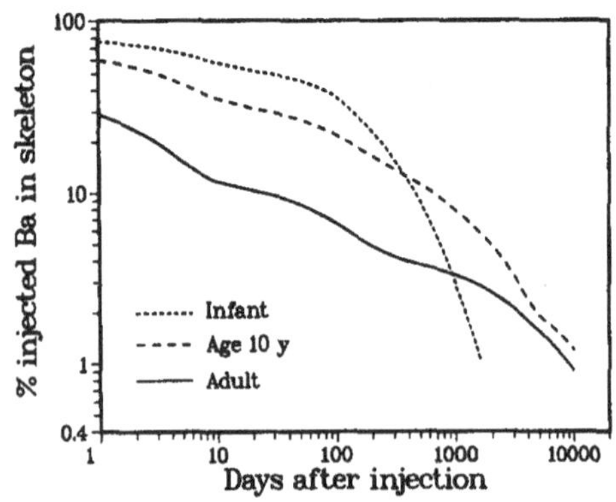

Fig. A7. Model predictions of the barium content of the skeleton as a function of age at injection and time after injection.

Table A-2 Age-specific transfer rates (d^{-1}) for barium model

	Age					
	3 mo	1 y	5 y	10 y	15 y	Adult
plasma to urinary bladder contents	0.747	1.64	1.79	1.31	0.777	2.24
plasma to ULI contents	6.72	14.78	16.13	11.83	6.99	20.16
plasma to trabecular bone surface	10.5	6.30	6.22	9.88	14.45	9.72
plasma to cortical bone surface	42.0	25.2	21.78	29.32	37.35	7.78
plasma to ST0	7.67	16.87	18.40	13.49	7.97	23.0
plasma to ST1	2.33	5.13	5.60	4.11	2.43	7.00
plasma to ST2	0.0466	0.103	0.112	0.0822	0.0486	0.140
bone surface to plasma	0.578	0.578	0.578	0.578	0.578	0.578
bone surface to exch[a] bone volume	0.116	0.116	0.116	0.116	0.116	0.116
exch bone volume to bone surface	0.0097	0.0097	0.0097	0.0097	0.0097	0.0097
exch bone volume to nonexch[a] volume	0.0042	0.0042	0.0042	0.0042	0.0042	0.0042
nonexch trabecular volume to plasma	0.00822	0.00288	0.00181	0.00132	0.000959	0.000493
nonexch cortical volume to plasma	0.00822	0.00288	0.00153	0.000904	0.000521	0.0000821
ST0 to plasma	2.56	5.62	6.13	4.50	2.66	7.67
ST1 to plasma	0.693	0.693	0.693	0.693	0.693	0.693
ST2 to plasma	0.00038	0.00038	0.00038	0.00038	0.00038	0.00038
f_1	0.6	0.3	0.3	0.3	0.3	0.2

[a]exch = exchangeable, nonexch = nonexchangeable

JAICRP 23-3/4-H

3. Strontium

(A25) Modifications of the model for application to strontium are based on the following assumed differences between strontium and the heavier alkaline earth elements:

1. The rate of removal of strontium from plasma is 15 day^{-1}, compared with 70 day^{-1} for radium and barium. This yields reasonable agreement with plasma retention data on strontium in humans, as summarized in *ICRP Publication 20* (ICRP, 1973) and Newton *et al.* (1990).
2. Fractional loss to urine plus faeces (that is, the fraction of activity leaving plasma that goes to urine via the kidneys and faeces via the gastrointestinal tract) is 0.15, or about half the value assigned to radium and barium (Harrison *et al.*, 1967; Newton *et al.*, 1990, 1991). The assumed ratio of cumulative urinary to faecal excretion is 3.3 (Spencer *et al.*, 1960; Fujita *et al.*, 1963; Cohn *et al.*, 1963; Harrison *et al.*, 1967; ICRP, 1973, 1988; Wenger and Soucas, 1975; Likhtarev *et al.*, 1975; Newton *et al.*, 1990), compared with ratios of about 0.028 for radium and 0.11 for barium.
3. It is assumed, in effect, that bone (as an organ) makes no distinction between the alkaline earth elements with regard to initial uptake from plasma but that strontium binds to non-exchangeable sites in bone crystal in preference to barium (and barium binds in preference to radium, as described earlier); however, the preference of bone crystal for strontium is assumed to be smaller in forming bone than existing bone. Specifically, it is assumed that: (a) 50% of strontium leaving exchangeable bone volume goes to non-exchangeable bone volume, compared with 30% for barium and 20% of radium; (b) the removal half-time of strontium from the exchangeable bone compartment is 80 days, compared with 50 days for barium and 30 days for radium; and (c) in children, fractional transfer from bone surfaces to exchangeable bone volume is 20% less for strontium than for the heavier alkaline earth elements due to a relatively high exchangeability of strontium incorporated into forming bone crystal. These values were chosen to yield reasonable agreement with skeletal and whole-body retention data for strontium in humans (Schulert *et al.*, 1959; Bishop *et al.*, 1960; Bauer *et al.*, 1961; Cohn *et al.*, 1963; Fujita *et al.*, 1963; Mays *et al.*, 1963; MacDonald *et al.*, 1965; Kereiakes *et al.*, 1966; Woodard and Dwyer, 1972; ICRP, 1973; Likhtarev *et al.*, 1975; Reeve and Hesp, 1976; Reeve *et al.*, 1976, 1983; Leggett *et al.*, 1982; Schlenker *et al.*, 1982; Newton *et al.*, 1990), *in vitro* data for strontium (Neuman *et al.*, 1963; Comar and Wasserman, 1964; Stark, 1968; McQueen and Smith, 1972), and comparative strontium, barium and radium data on humans and laboratory animals (Mays *et al.*, 1963; Bligh and Taylor, 1963; Harrison *et al.*, 1967; Domanski *et al.*, 1980). The basis for these assumptions is described in more detail elsewhere (Leggett, 1992).
4. Strontium is retained in soft tissues to a smaller extent than either barium or radium (Comar *et al.*, 1957; Schulert *et al.*, 1959; Mays *et al.*, 1963; ICRP, 1973; Schlenker *et al.*, 1982). Soft tissue compartment sizes and half-times for strontium are based on data of Comar *et al.* (1957), Schulert *et al.* (1959), and Mays *et al.* (1963) and the assumption that roughly 1% of total body strontium is in soft tissues after many years of chronic exposure (Schlenker *et al.*, 1982).

(A26) *Children.* As discussed in Section 5, the behaviour of strontium in children has been studied extensively. Figures A8 and A9 give the predicted soft tissue and skeleton content of strontium for infants, 10-year-old children and adults after entry into the systemic circulation based on the parameter values given in Table A-3.

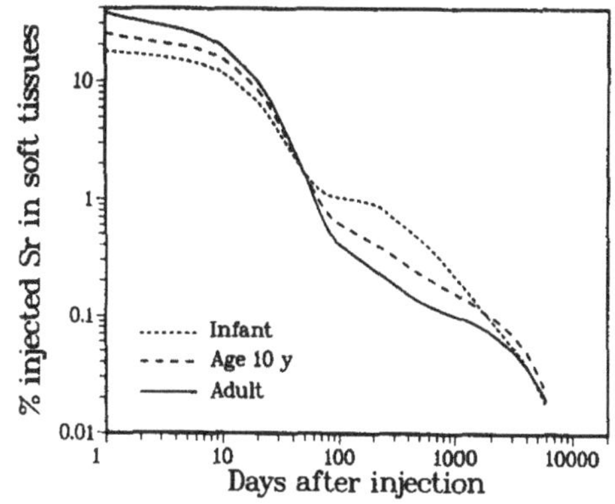

Fig. A8. Model predictions of the strontium content of the soft tissues as a function of age at injection and time after injection.

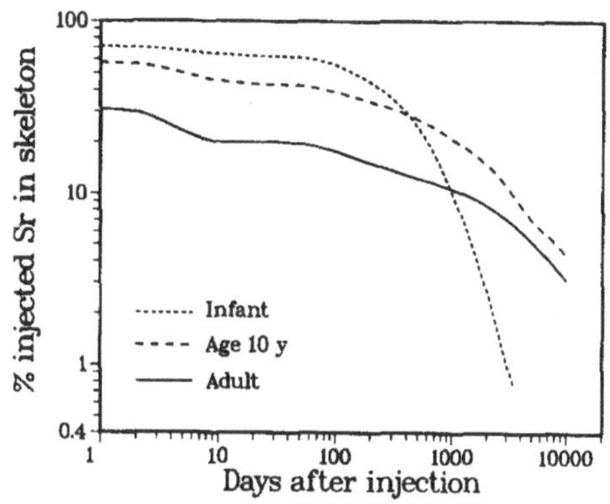

Fig. A9. Model predictions of the strontium content of the skeleton as a function of age at injection and time after injection.

Age-specific Biokinetic Model for Lead

(A27) The model is described in detail elsewhere (Leggett, 1993) and summarized here. The compartments and paths of movement used for lead differ slightly from those for the alkaline earth elements, due primarily to differences in the biological behaviour of lead and the alkaline earths but also because of differences in the type and amount of information available to derive parameter values for lead. The following changes are made for lead:

1. Red blood cells are considered as a separate compartment.
2. Excretion via sweat and hair (not shown in Fig. A1) is considered.
3. The liver is assumed to consist of two compartments, referred to as Liver 1 and Liver 2 (Fig. A1). Liver 1, which is assumed to have relatively short retention, is used primarily to model the high uptake of lead by the liver at early times and its relatively fast removal to plasma and the gastrointestinal tract (in bile). Liver 2, which is assumed to have relatively long retention, is used to help reproduce the blood-to-liver concentration ratios observed in persons chronically exposed to low levels of environmental lead.

Table A-3 Age-specific transfer rates (d^{-1}) for strontium model

	Age					
	3 mo	1 y	5 y	10 y	15 y	Adult
plasma to urinary bladder contents	0.577	1.27	1.38	1.02	0.600	1.73
plasma to ULI contents	0.175	0.385	0.42	0.308	0.182	0.525
plasma to trabecular bone surface	2.25	1.35	1.33	2.12	3.10	2.08
plasma to cortical bone surface	9.00	5.40	4.67	6.28	8.00	1.67
plasma to ST0	2.50	5.50	6.00	4.40	2.60	7.50
plasma to ST1	0.50	1.10	1.20	0.880	0.52	1.50
plasma to ST2	0.0010	0.0022	0.0024	0.0018	0.0010	0.0030
bone surface to plasma	0.601	0.601	0.601	0.601	0.601	0.578
bone surface to exch[a] bone volume	0.0924	0.0924	0.0924	0.0924	0.0924	0.116
exch bone volume to bone surface	0.0043	0.0043	0.0043	0.0043	0.0043	0.0043
exch bone volume to nonexch[a] volume	0.0043	0.0043	0.0043	0.0043	0.0043	0.0043
nonexch trabecular volume to plasma	0.00822	0.00288	0.00181	0.00132	0.000959	0.000493
nonexch cortical volume to plasma	0.00822	0.00288	0.00153	0.000904	0.000521	0.0000821
ST0 to plasma	0.833	1.83	2.00	1.47	0.867	2.50
ST1 to plasma	0.116	0.116	0.116	0.116	0.116	0.116
ST2 to plasma	0.00038	0.00038	0.00038	0.00038	0.00038	0.00038
f_1	0.6	0.4	0.4	0.4	0.4	0.3

[a]exch = exchangeable, nonexch = nonexchangeable

4. The kidneys are considered explicitly and are assumed to consist of two compartments, one with relatively short retention ("urinary path" in Fig. A1) and one with relatively long retention ("other kidney tissue" in Fig. A1). The "urinary path", which is assumed to lose activity to the urinary bladder contents, is used to model delayed urinary excretion of lead. The "other kidney tissue", which is assumed to return activity to plasma, is used to help reproduce blood-to-kidney concentration ratios observed in persons chronically exposed to lead.

Selection of parameter values for blood

(A28) Parameter values for the model (Table A-4) were based on: (1) experimental studies on humans receiving stable or radioactive lead by injection, acute inhalation or acute ingestion; (2) long-term balance studies; (3) autopsy measurements on environmentally exposed humans; (4) bioassay data and autopsy measurements on occupationally exposed persons; (5) injection data on laboratory animals, mainly baboons and beagles; and (6) considerations of changes with age in bone growth and remodelling. For application of the relatively plentiful but widely scattered data on stable lead in human tissues, it was useful to derive "most typical" (i.e. reference) distributions of lead in humans of different ages, based on comparison of these data. The details of evaluation of studies, reduction of data and selection of reference values are described elsewhere (Leggett, 1993).

(A29) *Adults.* At 1 day after introduction of radioactive lead into adult humans by injection or inhalation, the blood contains 40–75% (mean $58 \pm 12\%$) of the amount reaching the circulation (Hursh and Suomela, 1968; Booker *et al.*, 1969; Hursh *et al.*, 1969; Wells *et al.*, 1975; Chamberlain *et al.*, 1978; Morrow *et al.*, 1980; Heard and Chamberlain, 1984). Over the next few weeks, lead is lost from the blood with a net half-time on the order of 15–20 days (Rabinowitz *et al.*, 1973, 1974, 1976; Wells *et al.*, 1975; Heard and Chamberlain, 1984). Within a few hours after injection, 99% or more of blood lead is bound in or on red blood cells (Hursh *et al.*, 1969; Booker *et al.*, 1969; Wells *et al.*, 1975; Chamberlain *et al.*, 1978; Everson and Patterson, 1980; DeSilva, 1981; Manton and Cook, 1984; Heard and Chamberlain, 1984).

(A30) Immediately after injection of radioactive lead into humans, most of the injected activity disappears from blood at a rate of 1 min^{-1} or greater (Wells *et al.*, 1975; Chamberlain *et al.*, 1978). Increased activity in blood is then seen for a period of several hours, indicating return from extravascular spaces (Booker *et al.*, 1969; Wells *et al.*, 1975; Chamberlain *et al.*, 1978). In the present model, this early, rapid exchange between plasma and extravascular fluids is ignored. As is the case for radium and barium, it is assumed that lead leaves plasma at a rate of 70 day^{-1} and that a substantial portion of activity leaving plasma goes to a rapid-turnover soft tissue compartment that is three times as large as the plasma compartment. Inflow and outflow rates selected for red blood cells yield an estimate of roughly 58% of injected lead in red blood cells at 1–2 days after injection. The flow rate from red blood cells to plasma is set to yield a net half-time in blood of about 20 days in the time between a few days and a few weeks after injection, based on data for humans.

(A31) Data on injected lead in human subjects (Heard and Chamberlain, 1984), baboons (Cohen *et al.*, 1970), and beagles (Lloyd *et al.*, 1975) indicate that the liver rapidly accumulates 10–15% of systemic lead. Most of this is removed to plasma and the gastrointestinal tract over the first several weeks after injection, but a long-term retention component may also be present. In humans environmentally or occupationally

Table A-4 Age-specific transfer rates (d^{-1}) for lead model

	Age					
	3 mo	1 y	5 y	10 y	15 y	Adult
plasma to red blood cells (RBC)	20.0	24.8	25.60	23.04	20.16	28.0
plasma to urinary bladder contents	1.25	1.55	1.60	1.44	1.26	1.75
plasma to large intestine contents	0.50	0.62	0.64	0.576	0.504	0.70
plasma to trabecular bone surface	5.25	3.15	3.11	4.94	7.23	4.86
plasma to cortical bone surface	21.0	12.6	10.89	14.66	18.67	3.89
plasma to liver 1	3.50	4.34	4.48	4.03	3.53	4.90
plasma to kidneys (urinary path)	1.75	2.17	2.24	2.02	1.76	2.45
plasma to other kidney tissues	0.0175	0.0217	0.0224	0.0202	0.0176	0.0245
plasma to ST0	15.83	19.63	20.26	18.23	15.96	22.16
plasma to ST1	0.50	0.62	0.64	0.576	0.504	0.70
plasma to ST2	0.10	0.124	0.128	0.115	0.101	0.14
plasma to sweat	0.30	0.372	0.384	0.346	0.302	0.42
RBC to plasma	0.139	0.139	0.139	0.139	0.139	0.139
bone surface to plasma	0.65	0.65	0.65	0.65	0.65	0.50
bone surface to exch[a] bone volume	0.35	0.35	0.35	0.35	0.35	0.50
exch bone volume to bone surface	0.0185	0.0185	0.0185	0.0185	0.0185	0.0185
exch bone volume to nonexch[a] volume	0.0046	0.0046	0.0046	0.0046	0.0046	0.0046
nonexch trabecular volume to plasma	0.00822	0.00288	0.00181	0.00132	0.000959	0.000493
nonexch cortical volume to plasma	0.00822	0.00288	0.00153	0.000904	0.000521	0.0000821
liver 1 to plasma	0.0312	0.0312	0.0312	0.0312	0.0312	0.0312
liver 1 to small intestine contents	0.0312	0.0312	0.0312	0.0312	0.0312	0.0312
liver 1 to liver 2	0.00693	0.00693	0.00693	0.00693	0.00693	0.00693
liver 2 to plasma	0.00693	0.00693	0.00693	0.0019	0.0019	0.0019
kidneys (urinary path) to urine	0.139	0.139	0.139	0.139	0.139	0.139
other kidney tissue to plasma	0.00693	0.00693	0.00693	0.0019	0.0019	0.0019
ST0 to plasma	5.28	6.54	6.75	6.08	5.32	7.39
ST1 to plasma	0.00416	0.00416	0.00416	0.00416	0.00416	0.00416
ST1 to excreta[b]	0.00277	0.00277	0.00277	0.00277	0.00277	0.00277
ST2 to plasma	0.00038	0.00038	0.00038	0.00038	0.00038	0.00038
f_1	0.6	0.4	0.4	0.4	0.4	0.2

[a]exch = exchangeable, nonexch = nonexchangeable

[b]Mainly hair, but includes loss in nails and desquamated skin, for example.

exposed to lead, the ratio of the lead concentration in blood to that in liver typically is near 0.2 (Blanchard and Moore, 1970, 1971; Hamilton *et al.*, 1972; ICRP, 1975; Gross *et al.*, 1975; Barry, 1975, 1981; Iyengar, 1985).

(A32) Estimates of secretion of lead in human bile are highly variable. For example, data of Rabinowitz *et al.* (1976) indicate that biliary secretion represents no more than half of all endogenous secretion of lead into the gastrointestinal tract, while data of Ishihara and Matsushiro (1986) suggest that hepatic bile is the main route of elimination of absorbed lead from the body.

(A33) In this model, activity entering the liver is assumed to deposit in Liver 1. A small portion of the activity leaving Liver 1 is assigned to Liver 2, but most is divided between plasma and the small intestine. Removal rates from plasma to Liver 1 and from Liver 1 to plasma and to the small intestine (in hepatic bile) were chosen to yield reasonable agreement with data for humans and laboratory animals. Specifically, it is assumed that the removal half-time from Liver 1 is 10 days, that 10% of activity leaving Liver 1 deposits in Liver 2, and that the remaining 90% of outflow from Liver 1 is evenly divided between plasma and the contents of the small intestine.

(A34) The model for the kidneys is based on injection data for baboons (Cohen *et al.*, 1970) and beagles (Lloyd *et al.*, 1975), acute ingestion data for monkeys (Willes *et al.*, 1977) and autopsy data for environmentally exposed humans (Leggett, 1993). Parameters for the short-term compartment were based on data for beagles and non-human primates and those for the long-term compartment were set to yield agreement with observed relative liver, kidney and blood concentrations in chronically exposed humans. Lead that may be filtered by the kidneys but quickly reabsorbed to blood (Vander *et al.*, 1977; Victery *et al.*, 1979a,b) is ignored.

(A35) In addition to the rapidly exchanging soft tissue pool described in the discussion of the early circulation of lead, two other non-specific soft tissue pools (an intermediate-term pool and a long-term pool) are considered. Both of these are assumed to exchange activity with plasma. In addition, the intermediate-term pool is assumed to be the source of lead lost from the body in hair. Transfer rates representing exchange of the intermediate- and long-term soft tissue pools with plasma were based on autopsy data for chronically exposed humans, experimental data for humans ingesting lead over a period of months (Rabinowitz *et al.*, 1976), and injection data for beagles (Lloyd *et al.*, 1975) and baboons (Cohen *et al.*, 1970). The rate of movement of lead from the intermediate-term pool to hair was based on data of Kopito *et al.* (1967), Hammer *et al.* (1971), Roberts *et al.* (1974), Rabinowitz *et al.* (1976), Kijewski and Lowitz (1982) and Paschal *et al.* (1989) for human subjects.

(A36) Skeletal behaviour of lead appears to be qualitatively similar to that of the alkaline earth elements and quantitatively similar to that of barium or radium, if account is taken of the slower deposition of lead in the skeleton due to competition from red blood cells (Hursh, 1973; Lloyd *et al.*, 1975; Domanski and Trojanowska, 1980; Heard and Chamberlain, 1984). Lead has been used frequently as a marker of bone growth and osteon formation, and a close resemblance to calcium has been demonstrated in such studies (Vincent, 1957; Lacroix, 1960; Scheiman-Tagger and Brodie, 1964; Hong *et al.*, 1968; Yen and Shaw, 1977). Lead is incorporated into the crystalline structure of bone, where it replaces calcium ions (MacDonald *et al.*, 1951; Verbeeck *et al.*, 1981; Bres *et al.*, 1986; Miyake *et al.*, 1986). Autoradiographs of bone sections from baboons injected with ^{210}Pb indicate that a portion of skeletal activity remains near bone surfaces at 1 to 2 months after administration, as appears to be the case for radium and barium (see earlier). It also has been shown for some human subjects that the distribution of lead in bone is skewed toward bone surfaces for at least a few months after exposure, but the subjects generally have been exposed to heavy levels of lead that could affect bone metabolism (Lindh *et al.*, 1978; Flood *et al.*, 1988). Burial of lead beneath the surfaces in regions of bone formation has been observed, and there is evidence that lead is eventually distributed throughout the bone volume (Vincent, 1957; Lacroix, 1960; Scheiman-Tagger and Brodie, 1964; Hong *et al.*, 1968; Yen and Shaw, 1977; Lindh *et al.*, 1978; Hu *et al.*, 1989). In beagles, long-term skeletal retention of lead is similar to that of strontium and radium (Hursh, 1973; Lloyd *et al.*, 1975). Hursh (1973) noted that, inasmuch as lead, strontium and radium are all incorporated into the bone crystal, it might be expected that long-term skeletal losses would be largely controlled by the rate of bone resorption.

(A37) In a study of the comparative behaviours of injected lead, calcium and barium in bone of rabbits, Domanski and Trojanowska (1980) found that the build-up of lead in the skeleton is similar to that of barium and greater than that of calcium when related to integrated activity in plasma. Similar results for lead and calcium were obtained by Heard and Chamberlain (1984) for humans injected with isotopes of these two elements. In baboons (Cohen *et al.*, 1970) and humans (Heard and Chamberlain, 1984), there was evidence of rapid skeletal uptake of roughly 10% of injected lead, followed in some cases by some loss over the first day or two, and then continual uptake for an extended period as activity returned from red blood cells and soft tissue to plasma. By 20 days after injection of lead, the human skeleton may contain roughly 20% of the injected amount (Hard and Chamberlain, 1984). This relatively low uptake by the skeleton at early times compared with radium, for example, apparently reflects a competition with red blood cells not experienced by the alkaline earth elements. The later build-up in the skeleton results from the gradual release of activity from red blood cells and the relatively longer retention of lead in the skeleton than in red blood cells. It would appear reasonable to apply the same bone models to radium and lead, provided account is taken of differences in the rate of uptake of these two elements by bone due to differences in their affinities for red blood cells and soft tissues, and provided appropriate changes are made in parameter values describing early retention on bone surfaces.

(A38) The early pattern of skeletal retention of lead observed in adult humans (Heard and Chamberlain, 1984) and baboons (Cohen *et al.*, 1970) can be reproduced reasonably well by assuming that: (1) fractional deposition of lead on trabecular or cortical bone surfaces (expressed as a fraction of activity leaving plasma) is one-half the corresponding value for radium; (2) lead is removed from bone surfaces at a rate of 1 day^{-1}; and (3) half the lead leaving bone surfaces returns to plasma, with the other half migrating to exchangeable bone volume.

(A39) Typically, 3–5% of injected or absorbed lead is lost in urinary excretion during the first day (Hursh and Suomela, 1968; Hursh *et al.*, 1969; Booker *et al.*, 1969; Hursh and Mercer, 1970; Chamberlain *et al.*, 1978). Gradual loss of lead from red blood cells, liver, kidneys and other soft tissues over the first few weeks can be accounted for by a slow loss in urine and faeces (about 30% of systemic activity in the first 20 days) and a continual increase in skeletal lead (Heard and Chamberlain, 1984). The urinary to faecal excretion ratio is roughly 2 during days 3–14 after absorption of lead to blood in humans (Wells *et al.*, 1975; Chamberlain *et al.*, 1978; Heard and Chamberlain, 1984). Rates of clearance from plasma to urine and to the gastrointestinal tract were based on these data and the assumed rate of removal from Liver 1 to the contents of the small intestine (see earlier discussion of the liver). Activity moving directly from plasma to the gastrointestinal tract contents is assumed to deposit in the contents of the upper large intestine and hence is assumed not to be available for reabsorption to plasma. Lead in sweat is assumed to be lost directly from plasma. The rate of movement of lead from plasma to sweat was based on findings of Shiels (1954), Hohnadel *et al.* (1973), Rabinowitz *et al.* (1976) and Stauber and Florence (1988) for human subjects.

(A40) *Children.* Data for non-human primates (Willes *et al.*, 1977; Pounds *et al.*, 1978; Kneip *et al.*, 1983) and rodents (Momcilovic and Kostial, 1974; Kello *et al.*, 1975; Jugo *et al.*, 1975; Kostial *et al.*, 1978; Jugo, 1980; Keller and Doherty, 1980) indicate that early retention of lead is greater in growing than mature animals and that much of the variation with age is due to elevated uptake and/or retention of lead by the immature skeleton. In an ingestion experiment with rhesus monkeys, absorbed lead was excreted at a lower rate by infants than by adults during the first few weeks after administration (Pounds *et al.*, 1978). Some differences in uptake and/or retention of lead by the brain, liver and kidneys have been observed between immature and mature rodents (Momcilovic and Kostial, 1974; Keller and Doherty, 1980). The combined age-specific data for laboratory animals and environmentally exposed humans are reasonably consistent with assumptions of higher uptake and faster release of lead by the immature than the mature skeleton and substantial retention of lead by red blood cells and soft tissues at all ages. Also, autopsy data for environmentally exposed humans indicate potentially greater uptake and/or retention of lead in some soft tissues of children than in adults and potentially lower uptake and/or retention in the liver and kidneys in children (Barry, 1975, 1981; Leggett, 1993). Thus, estimated ratios of total contents of blood, liver, kidneys and other soft tissues change considerably with age, particularly when account is taken of changes with age in the relative masses of the different organs and tissues (Leggett, 1993).

(A41) Autopsy data for persons chronically exposed to environmental lead suggest that the skeletal content of lead increases throughout life (Leggett, 1993). These data also indicate that the skeleton contains roughly one-third of total-body lead during infancy, one-half to two-thirds of total-body lead in young children and teenagers and 90% or more of total-body lead in middle-aged and older persons (Barry, 1975, 1981; Leggett, 1993).

(A42) The following modifications of the lead model for adults are made for application to children.

1. Deposition rates on trabecular and cortical bone surfaces are assumed to be proportional to age-specific values estimated for the alkaline earth elements as given earlier, using as a reference point the fraction of plasma lead estimated to deposit on all bone surfaces in mature adults.
2. As in the models for the alkaline earth elements, deposition of lead in soft tissues and excreta is assumed to be reduced in children due to elevated uptake of lead by bone. The method of derivation of deposition fractions for children is the same as described earlier for radium.
3. For general agreement with autopsy data on environmentally exposed humans, retention times in the long-term compartments of liver and kidneys are assumed to be lower in young children than in older children and adults. Specifically, removal

half-times from Liver 2 to plasma and from "other kidney tissue" to plasma are each assumed to be 100 days at age 5 years or less, compared with assumed removal half-times from these compartments of 1 year in older children and adults.

4. As in the model for strontium, fractional transfer from bone surfaces to exchangeable bone volume is assumed to be smaller (by 30% in this case) during ages 0–15 years than during adulthood. This assumption is needed to reproduce autopsy data on children exposed to environmental lead (Leggett, 1993).

(A43) Figures A10 and A11 give estimated lead contents of soft tissues and skeleton of infants, 10-year-old children, and adults as a function of time after injection, based on the parameter values given in Table A-4.

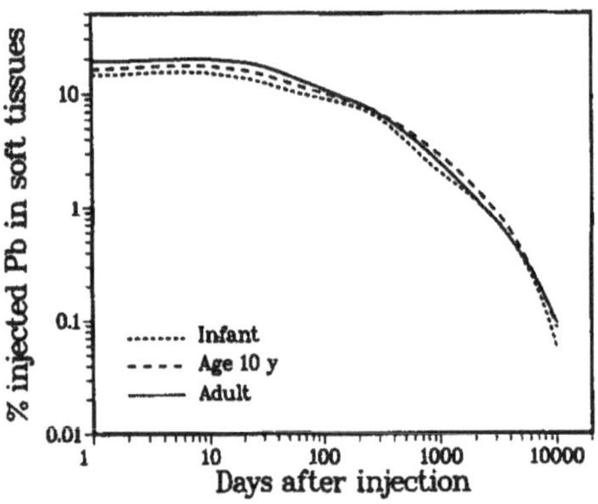

Fig. A10. Model predications of the lead content of the soft tissues as a function of age at injection and time after injection.

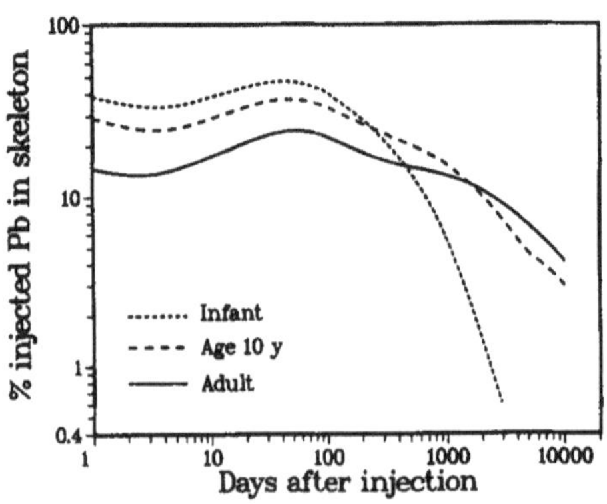

Fig. A11. Model predictions of the lead content of the skeleton as a function of age at injection and time after injection.

References

Atherton, D. R., Stover, B. J. and Mays, C. W. (1965) Soft tissue retention of Ra-226 in the beagle. *Health Phys.* **11**, 101–108.

Barnes, O. W., Bishop, M., Harrison, G. E. and Sutton, A. (1961) Comparison of the plasma concentration and urinary excreton of strontium and calcium in man. *Int. J. Radiat. Biol.* **3**, 637–646.

Barry, P. S. I. (1975) A comparison of concentrations of lead in human tissues. *Br. J. Ind. Med.* **32**, 119–139.

Barry, P. S. I. (1981) Concentrations of lead in tissues of children. *Br. J. Ind. Med.* **38**, 61–71.

Bauer, G. C. H., Carlsson, A. and Lindquist, B. (1955) A comparative study on the metabolism of ^{90}Sr and ^{45}Ca. *Acta Physiol. Scand.* **35**, 56–66.

Bauer, G. C. H., Carlsson, A. and Lindquist, B. (1957) Metabolism of Ba-140 in man. *Acta Orth. Scand.* **26**, 241–257.

Bauer, G. C. H., Carlsson, A. and Lindquist, B. (1961) Metabolism and homeostatic function of bone. In: *Mineral Metabolism,* 1, Part B (C. Comer and L. Bronner, Eds), pp. 609–676. Academic Press, New York.

Bishop, M., Harrison, G. E., Raymond, W. H. A., Sutton, A. and Rundo, J. (1960) Excretion and retention of radioactive strontium in normal man following a single intravenous injection. *Int. J. Radiat. Biol.* **2**, 125–142.

Blanchard, R. L. and Moore, J. B. (1970) Pb-210 and Po-210 in tissues of some Alaskan residents as related to consumption of caribou or reindeer meat. *Health Phys.* **18**, 127–134.

Blanchard, R. L. and Moore, J. B. (1971) Body burden, distribution and internal dose of Pb-210 and Po-210 in a uranium miner population. *Health Phys.* **21**, 499–518.

Bligh, P. H. and Taylor, D. M. (1963) Comparative studies of the metabolism of strontium and barium in the rat. *Biochem. J.* **87**, 612–618.

Booker, D. V., Chamberlain, A. C., Newton, D. and Stott, A. N. B. (1969) Uptake of radioactive lead following inhalation and injection. *Br. J. Radiol.* **42**, 457–466.

Bres, E. F., Voegel, J. C., Barry, J. C., Waddington, W. G. and Frank, R. M. (1986) Feasibility study for the detection of lead substitution sites in the hydroxyapatite crystal structure using high-resolution electron microscopy (HREM) at optimum focus. *J. Appl. Cryst.* **19**, 168–173.

Bruenger, F. W. and Lloyd, R. D. (1989) The influence of age at time of exposure to Ra-226 or Pu-239 on distribution, retention, post-injection survival and bone tumor induction in beagle dogs. In: *Thirty-fourth Annual Meeting of the Health Physics Society, Abstracts of Papers Presented at the Meeting, June 25–29, 1989, Albuquerque Convention Center, Albuquerque, NM,* Vol. 56 (Suppl. 1), p. 27. Pergamon Press, New York.

Bruenger, F. W., Smith, J. M., Atherton, D. R., Jee, W. S. S., Lloyd, R. D. and Stevens, W. (1983) Skeletal retention and distribution of Ra-226 and Pu-239 in beagles injected at ages ranging from 2 days to 5 years. *Health Phys.* **44**, 513–527.

Bruenger, F. W., Lloyd, R. D. and Miller, S. C. (1989) The influence of age at time of exposure on the distribution and retention of Ra-226 or Pu-239 in beagle dogs. *Inhalation Toxicology Research Institute Annual Report,* 1988–1989, pp. 247–250.

Chamberlain, A. C., Heard, M. J., Little, P., Newton, D., Wells, A. C. and Wiffen, R. D. (1978) Investigations into lead from motor vehicles. Harwell: Environmental and Medical Sciences Division, AERE-R 9198.

Cohen, N., Eisenbud, M. and Wrenn, M. E. (1970) *The Retention and Distribution of Lead-210 in the Adult Baboon.* Progress Report. Radioactivity Studies, New York University Medical Center, New York.

Cohn, S. H., Lippincott, S. W., Gusmano, E. A. and Robertson, J. S. (1963) Comparative kinetics of ^{47}Ca and ^{85}Sr in man. *Radiat. Res.* **19**, 104–119.

Comar, C. L. and Wasserman, R. H. (1958) Strontium-calcium metabolism in man and animals as studied by radioisotope methods. In: *Radioisotopes in Scientific Research,* Vol. 4, pp. 191–206. Pergamon Press, New York.

Comar, C. L. and Wasserman, R. H. (1964) In: *Mineral Metabolism II,* Part A (C. L. Comar and L. Bronner, Eds), pp. 523–572. Academic Press, New York.

Comar, C. L., Wasserman, R. H., Ullberg, S. and Andrews, G. A. (1957) Strontium metabolism and strontium-calcium discrimination in man. *Proc. Soc. Exp. Biol. Med.* **95**, 386–391.

Cuddihy, R. G. and Griffith, W. C. (1972) A biological model describing tissue distribution and whole body retention of barium and lanthanum in beagle dogs after inhalation and gavage. *Health Phys.* **23**, 621–633.

Decker, C. F., Kaspar, L. V. and Norris, W. P. (1964) The variation of strontium metabolism with age in the dog. *Radiat. Res.* **23**, 475–490.

Della Rosa, R. J., Goldman, M. and Wolf, H. G. (1967) Uptake and retention of orally administered Ra-226 and Ba-133: A preliminary report. University of California at Davis, **472–114**, pp. 40–41.

DeSilva, P. E. (1981) Determination of lead in plasma and studies on its relationship to lead in erythrocytes. *Br. J. Ind. Med.* **38**, 209–217.

Doig, A. T. (1976) Baritosis: a benign pneumoconiosis. *Thorax* **31**, 30–39.

Domanski, T. and Trojanowska, B. (1980) Studies on metabolic kinetics of lead and alkaline earth elements (Ca, barium). *Acta Physiol. Pol.* **31**, 439–447.

Domanski, T., Witkowska, D. and Garlicka, I. (1980) Influence of age on the discrimination of barium in comparison with strontium during their incorporation into compact bone. *Acta Physiol. Pol.* **31**, 289–296.

Domanski, T., Liniecki, J. and Witkowska, D. (1969) Kinetics of calcium strontium, barium, and radium in rats. In: *Delayed Effects of Bone Seeking Radionuclides* (C. W. Mays, Ed.), pp. 79–94. University of Utah Press, Salt Lake City.

Ellsasser, J. C., Farnham, J. E. and Marshall, J. H. (1969) Comparative kinetics and autoradiography of Ca-45 and Ba-133 in ten-year-old beagle dogs. *J. Bone Jt. Surg.* **51A**, 1397–1412.

Erre, N., Manca, F. and Parodo, A. (1980) The short-term retention of barium in man. *Health Phys.* **38**, 225–227.

Everson, J. and Patterson, C. C. (1980) "Ultra-clean" isotope dilution/mass spectrometric analyses for lead in human blood plasma indicate that most reported values are artificially high. *Clin. Chem.* **26**, 1603–1607.

Farnham, J. E. and Rowland, R. E. (1965) The retention of Ba-133 in beagles. In: *Radiological Physics Division Annual Report, July 1964 through June 1965.* Argonne National Laboratory, **ANL-7060**, pp. 70–73.

Fisenne, I. M., Keller, H. W. and Harley, N. H. (1981) Worldwide measurement of Ra-226 in human bone: estimate of skeletal alpha dose. *Health Phys.* **40**, 163–171.

Flood, P. R., Schmidt, P. F., Wesenberg, G. R. and Gadeholt, H. (1988) The distribution of lead in human hemopoietic tissue and spongy bone after lead poisoning and Ca-EDTA chelation therapy. *Arch. Toxicol.* **62**, 295–300.

Fujita, M., Yabe, A., Ueno, K., Oshino, M. and Okuyama, N. (1963) The behavior of strontium-85 in a normal man following a single ingestion—absorption and excretion. *Health Phys.* **9**, 407–415.

Garner, R. J. (1960) Distribution of radioactive barium in eye tissues. *Nature* **184**, 733–734.

Gross, S. B., Pfitzer, E. A., Yeager, D. W. and Kehoe, R. A. (1975) Lead in human tissues. *Toxicol. Appl. Pharmacol.* **32**, 638–651.

Hallden, N. A., Fisenne, I. M. and Harley, J. H. (1963) Radium-226 in human diet and bone. *Science* **140**, 1327–1329.

Hamilton, E. I., Minski, M. J. and Cleary, J. J. (1972) The concentration and distribution of some stable elements in healthy human tissues from the United Kingdom. *Sci. Total Environ.* **1**, 341–374.

Hammer, D. I., Finklea, J. F., Hendricks, R. H., Hinners, T. A., Riggan, W. B. and Shy, C. (1971) Trace metals in human hair as a simple epidemiologic monitor of environmental exposure. In: *Proceedings of the 5th Conference on Trace Substances in Environmental Health* (D. D. Hemphill, Ed.), pp. 25–38.

Hardy, E., Rivera, J., Fisenne, I., Pond, W. and Hogue, D. (1969) Comparative utilization of dietary radium-226 and other alkaline earths by pigs and sheep. In: *Radiation Biology of the Fetal and Juvenile Mammal* (M. R. Sikov and D. D. Mahlum, Eds), pp. 183–190. Proceedings of the Ninth Annual Hanford Biology Symposium at Richland, Washington, May 5–8, 1969. U.S. AEC, Division of Technical Information.

Harrison, G. E. (1981) Whole body retention of the alkaline earths in adult man. *Health Phys.* **40**, 95–99.

Harrison, G. E., Carr, T. E. F. and Sutton, A. (1967) Distribution of radioactive calcium, strontium, barium and radium following intravenous injection into a healthy man. *Int. J. Radiat. Biol.* **13**, 235–247.

Heard, M. J. and Chamberlain, A. C. (1984) Uptake of lead by human skeleton and comparative metabolism of lead and alkaline earth elements. *Health Phys.* **47**, 857–865.

Hohnadel, D. C., Sunderman, F. W., Nechay, M. W. and McNeely, M. D. (1973) Atomic absorption spectrometry of nickel, copper, zinc, and lead in sweat collected from healthy subjects during sauna bathing. *Clin. Chem.* **19**, 1288–1292.

Hong, Y. C., Yen, P. K.-J. and Shaw, J. H. (1968) An analysis of the growth of the cranial vault in rabbits by vital staining with lead acetate. *Calc. Tiss. Res.* **2**, 271–285.

Hu, H., Milder, F. L. and Burger, D. E. (1989) X-ray fluorescence: Issues surrounding the application of a new tool for measuring burden of lead. *Environ. Res.* **49**, 295–317.

Hursh, J. B. (1973) Retention of Pb-210 in beagle dogs. *Health Phys.* **25**, 29–35.

Hursh, J. B. and Lovaas, A. (1961) The biliary excretion of radium. *Health Phys.* **6**, 183–188.

Hursh, J. B. and Lovaas, A. (1963) Radium-226 in bone and soft tissues of man. *Nature* **198**, 265–268.

Hursh, J. B. and Mercer, T. T. (1970) Measurement of Pb-212 loss rate from human lungs. *J. Appl. Physiol.* **28**(3), 268–274.

Hursh, J. B. and Suomela, J. (1968) Absorption of Pb-212 from the gastrointestinal tract of man. *Acta Radiol., Ther. Phys. Biol.* **7**, 108–120.

Hursh, J. B., Schraub, A., Sattler, E. L. and Hofmann, H. P. (1969) Fate of Pb-212 inhaled by human subjects. *Health Phys.* **16**, 257–267.

ICRP (1973) *Alkaline Earth Metabolism in Adult Man,* ICRP Publication 20. Pergamon Press, Oxford.

ICRP (1975) *Report of the Task Group on Reference Man,* ICRP Publication 23. Pergamon Press, Oxford.

ICRP (1988) *Individual Monitoring for Intakes of Radionuclides by Workers: Design and Interpretation,* ICRP Publication 54. Pergamon Press, Oxford.

ICRP (1989) *Age-dependent Doses to Members of the Public from Intake of Radionuclides: Part 1,* ICRP Publication 56. Pergamon Press, Oxford.

Ishihara, N. and Matsushiro, T. (1986) Biliary and urinary excretion of metals in humans. *Arch. Environ. Health* **41**, 324–330.

Iyengar, G. V. (1985) Concentrations of 15 trace elements in some selected adult human tissues and body fluids of clinical interest from several countries: Results from a pilot study for the establishment of reference values. Institute of Medicine, Juelich Nuclear Center, Federal Republic of Germany.

Jee, W. S. S. and Polig, E. (1989) Local distribution and dosimetry of ^{226}Ra in the beagle skeleton. In: *Thirty-fourth Annual Meeting of the Health Physics Society, Abstracts of Papers Presented at the Meeting, June 25–29, 1989, Albuquerque Convention Center, Albuquerque, NM*, Vol. 56 (Suppl. 1), p. 26. Pergamon Press, New York.

Jugo, S. (1980) Chelatable fraction of Pb-203 in blood of young and adult rats. *Environ. Res.* **21**, 336–342.

Jugo, S., Maljkovic, T. and Kostial, K. (1975) The effect of chelating agents on lead excretion in rats in relation to age. *Environ. Res.* **10**, 271–279.

Keane, A. T. and Schlenker, R. A. (1987) Long-term loss of radium in 63 subjects exposed at ages 6 to 46. In: *Age-related Factors in Radionuclide Metabolism and Dosimetry* (G. B. Gerber, H. Métivier and H. Smith, Eds), pp. 127–135. Martinus Nijhoff Publishers, Dordrecht.

Keller, C. A. and Doherty, R. A. (1980) Distribution and excretion of lead in young and adult female mice. *Environ. Res.* **21**, 217–228.

Kello, D., Kostial, K. and Harrison, G. E. (1975) Influence of age and temperature of incineration on the loss of lead from rat bone. *Health Phys.* **28**, 169–170.

Kereiakes, J. G., Wellman, H. N. and Saenger, E. L. (1966) Radiation Exposure from Radiopharmaceuticals in Children. In: *Proc. First Int. Cong. IRPA* (W. S. Snyder, Ed.), pp. 775–781.

Kijewski, H. and Lowitz, H. D. (1982) The estimation of lead in hydride form from biopsy specimens from the iliac crests of patients who had contracted lead poisoning many years previously. *Arch. Toxicol.* **50**, 301–311.

Kneip, T. J., Mallon, R. P. and Harley, N. H. (1983) Biokinetic modelling for mammalian lead metabolism. *Neurotoxicology* **4**(3), 189–192.

Kopito, L., Byers, R. K. and Shwachman, H. (1967) Lead in hair of children with chronic lead poisoning. *New Engl. J. Med.* **276**, 949–953.

Korsunskii, V. N., Tarasov, N. F. and Naumenko, A. Z. (1981) Clinical evaluation of Ba-133m as an osteotropic agent. ORNL/TR-86/30 (7pgs). Translated from the Russian UDC 616.71-006-073.916. *Medit. Radiol.* **10**, 45–48.

Kostial, K., Kello, D., Jugo, S., Rabar, I. and Maljkovic, T. (1978) Influence of age on metal metabolism and toxicity. *Environ. Health Perspect.* **25**, 81–86.

Lacroix, P. (1960) Ca-45 autoradiography in the study of bone tissue. In: *Bone as a Tissue* (K. Rodahl, J. T. Nicholson and E. M. Brown, Jr., Eds), pp. 262–279. McGraw-Hill, New York.

Leggett, R. W. (1992) A generic age-specific biokinetic model for calcium-like elements. *Radiat. Prot. Dosim.* **41**, 183–198.

Leggett, R. W. (1993) An age-specific kinetic model of lead metabolism in humans. *Environ. Health Perspect.* **101**, 598–616.

Leggett, R. W., Eckerman, K. F. and Williams, L. R. (1982) Strontium-90 in bone: a case study in age-dependent dosimetric modeling. *Health Phys.* **43**, 307–322.

Leggett, R. W., Eckerman, K. F., Dunning, D. E., Cristy, M., Crawford-Brown, D. J. and Williams, L. R. (1984) On estimating dose rates to organs as a function of age following internal exposure to radionuclides. ORNL/TM-8265.

LeRoy, G. V., Rust, J. H. and Hasterlik, R. J. (1966) The consequences of ingestion by man of real and simulated fallout. *Health Phys.* **12**, 449–473.

Likhtarev, I. A., Dobroskok, I. A., Ilyin, L. A., Krasnoschekova, G. P., Likhtareva, T. M., Smirnov, B. I., Sobolev, E. P., Shamov, V. P. and Shapiro, E. L. (1975) A study of certain characteristics of strontium metabolism in a homogeneous group of human subjects. *Health Phys.* **28**, 49–60.

Lindh, U., Brune, D. and Nordberg, G. (1978) Microprobe analysis of lead in human femur by proton induced X-ray emission (PIXE). *Sci. Total Environ.* **10**, 31–37.

Liniecki, J. (1971) Kinetics of calcium, strontium, barium and radium in rabbits. *Health Phys.* **21**, 367–376.

Lloyd, R. D., Mays, C. W., Atherton, D. R. and Bruenger, F. W. (1975) Pb-210 studies in beagles. *Health Phys.* **28**, 575–583.

Lloyd, R. D., Mays, C. W. and Atherton, D. R. (1976a) Distribution of injected Ra-226 and Sr-90 in the beagle skeleton. *Health Phys.* **30**, 183–189.

Lloyd, R. D., Mays, C. W., Atherton, D. R., Taylor, G. N. and Van Dilla, M. A. (1976b) Retention and skeletal dosimetry of injected Ra-226, Ra-228, and Sr-90 in beagles. *Radiat. Res.* **66**, 274–287.

Lloyd, R. D., Mays, C. W., Taylor, G. N., Atherton, D. R., Bruenger, F. W. and Jones, C. W. (1982) Radium-224 retention, distribution, and dosimetry in beagles. *Radiat. Res.* **92**, 280–295.

Lloyd, R. D., Bruenger, F. W., Jones, C. W., Taylor, G. N. and Mays, C. W. (1983a) Radium retention in mature beagles injected at 5 years of age. *Radiat. Res.* **94**, 210–216.

Lloyd, R. D., Bruenger, F. W., Mays, C. W. and Jones, C. W. (1983b) Skeletal radon-to-radium ratios in neonatal, juvenile and mature beagles and in adult St. Bernards. *Health. Phys.* **44**, 61–63.

Lloyd, R. D., Jones, C. W., Bruenger, F. W., Atherton, D. R. and Mays, C. W. (1983c) Radium retention and dosimetry in juvenile beagles. *Radiat. Res.* **94**, 295–304.

Lloyd, R. D., Taylor, G. N., Jones, C. W. and Mays, C. W. (1983d) Radium retention and dosimetry in the St. Bernard. *Radiat. Res.* **95**, 150–157.

Loutit, J. F. and Russell, R. S. Eds (1961) *Progress in Nuclear Energy, Series VI, Biological Sciences. Vol. 3, The Entry of Fission Products into Food Chains.* Pergamon Press, New York.

MacDonald, N. S., Ezmirlian, F., Spain, P. and McArthur, C. (1951) The ultimate site of skeletal deposition of strontium and lead. *J. Biol. Chem.* **189**, 387–399.

MacDonald, N. S., Figueroa, W. G. and Urist, M. R. (1965) Short-term retention of strontium-85 and estimation of initial strontium-90 burdens in humans. *Health Phys.* **11**, 1187–1194.

Maletskos, C. J., Keane, A. T., Telles, N. C. and Evans, R. D. (1969) Retention and absorption of Ra-224 and Th-234 and some dosimetric considerations of Ra-224 in human beings. In: *Delayed Effects of Bone-seeking Radionuclides* (C. W. Mays, W. S. S. Jee, R. D. Lloyd, B. J. Stover, J. H. Dougherty and G. N. Taylor, Eds), pp. 29–49. University of Utah Press, Salt Lake City, UT.

Manton, W. I. and Cook, J. D. (1984) High accuracy (stable isotope dilution) measurements of lead in serum and cerebrospinal fluid. *Br. J. Ind. Med.* **41**, 313–319.

Mays, C. W., Christensen, W. R., Lloyd, R. D., Atherton, D. R., Parmley, W. W. and Pitchford, G. S. (1962) Radium studies in a man soon after injection. In: *Research in Radiobiology*, University of Utah, COO-226, pp. 71–77.

Mays, C. W., Lloyd, R. D., Christensen, W. R., Atherton, D. R. and Pitchford, G. S. (1963) Radium metabolism in a man. *Radiat. Res.* **19**, 210 (abstract).

McQueen, C. M. and Smith, D. A. (1972) The relative uptake of ^{45}Ca and ^{85}Sr into bone. In: *Second Int. Conf. on Strontium Metabolism*, Glasgow and Strontian, 16–19 Aug. CONF-720818, pp. 403–411.

Miller, C. E. and Finkel, A. J. (1965) A reexamination of retention patterns in patients who received radium by multiple injections 33 years earlier. In: *Health Division Gamma-ray Spectroscopy Group Annual Report, July 1964 through June 1965*, Argonne National Laboratory, **ANL-7217**, pp. 5–90.

Miyake, M., Ishigaki, K. and Suzuki, T. (1986) Structure refinements of Pb^{2+} ion-exchanged apatites by X-ray powder pattern-fitting. *J. Solid State Chem.* **61**, 230–235.

Momcilovic, B. and Kostial, K. (1974) Kinetics of lead retention and distribution in suckling and adult rats. *Environ. Res.* **8**, 214–220.

Morrow, P. E., Beiter, H., Amato, F. and Gibb, F. R. (1980) Pulmonary retention of lead: An experimental study in man. *Environ. Res.* **21**, 373–384.

Müller, W. A. (1987) Age related retention and dose burden after injection. In: *Age-related Factors in Radionuclide Metabolism and Dosimetry* (G. B. Gerber, H. Métivier and H. Smith, Eds), pp. 26–28. Martinus Nijhoff Publishers, Dordrecht.

Muth, H. and Glöbel, B. (1983) Age dependent concentration of Ra-226 in human bone and some transfer factors from diet to human tissues. *Health Phys.* **44**, 113–121.

Neuman, W. F., Bjornerstedt, R. and Mulryan, B. J. (1963) Synthetic hydroxy apatite crystals II. Aging and strontium incorporation. *Arch. Biochem. Biophys.* **101**, 215–224.

Newton, D., Rundo, J. and Harrison, G. E. (1977) The retention of alkaline earth elements in man, with special reference to barium. *Health Phys.* **33**, 45–53.

Newton, D., Harrison, G. E., Rundo, J., Kang, C. and Warner, A. J. (1990) Metabolism of Ca and Sr in late adult life. *Health Phys.* **59**, 433–442.

Newton, D., Harrison, G. E., Kang, C. and Warner, A. J. (1991) The metabolism of injected barium in six healthy men. *Health Phys.* **61**, 191–201.

Norris, W. P., Speckman, T. W. and Gustafson, P. F. (1955) Studies on the metabolism of radium in man. *Am. J. Roentgenol., Radium Ther. Nucl. Med.* **73**, 785–802.

Parks, N. J. and Keane, A. T. (1983) Consideration of age-dependent radium retention in people on the basis of the beagle model. *Health Phys.* **44**, 103–112.

Parks, N. J., Pool, R. R., Williams, J. R. and Wolf, H. G. (1978) Age and dosage-level dependence of radium retention in beagles. *Radiat. Res.* **75**, 617–632.

Paschal, D. C., DiPietro, E. S., Phillips, D. L. and Gunter, E. W. (1989) Age dependence of metals in hair in a selected U.S. population. *Environ. Res.* **48**, 17–28.

Pounds, J. G., Marlar, R. J. and Allen, J. R. (1978) Metabolism of lead-210 in juvenile and adult rhesus monkeys (*Macaca mulatta*). *Bull. Environ. Contam. Toxicol.* **19**, 684–691.

Qiyue, H., Zhonghon, Z. and Shijun, L. (1988) Radium-226: Concentration and distribution in human bone (abstract). *Chin. J. Radiol. Med. Prot.* **8**, 224–225.

Rabinowitz, M. B., Wetherill, G. W. and Kopple, J. D. (1973) Lead metabolism in the normal human: Stable isotope studies. *Science* **182**, 725–727.

Rabinowitz, M. B., Wetherill, G. W. and Kopple, J. D. (1974) Studies of human lead metabolism by use of stable isotope tracers. *Environ. Health Perspect.* **5-7**, 145–153.

Rabinowitz, M. B., Wetherill, G. W. and Kopple, J. D. (1976) Kinetic analysis of lead metabolism in healthy humans. *J. Clin. Invest.* **58**, 260–270.

Rajewsky, B., Belloch-Zimmermann, V., Loehr, E. and Stahlhofen, W. (1965) Ra-226 in human embryonic tissue, relationship of activity to the stage of pregnancy, measurement of natural Ra-226 occurrence in the human placenta. *Health Phys.* **11**, 161–169.

Reeve, J. and Hesp, R. (1976) A model-independent comparison of the rates of uptake and short term retention of Ca-47 and Sr-85 by the skeleton. *Calcif. Tissue Res.* **22**, 183–189.

Reeve, J., Wotton, R. and Hesp, R. (1976) A new method for calculating the accretion rate of bone calcium and some observations on the suitability of strontium-85 as a tracer for bone calcium. *Calcif. Tissue Res.* **20**, 121–135.

Reeve, J., Green, J. R., Maletskos, C. J. and Neer, R. M. (1983) Skeletal retention of Ca-45 and Sr-85 compared: Further studies on intravenously injected Sr-85 as a tracer for skeletal calcium. *Calcif. Tissue Int.* **35**, 9–15.

Richmond, C. R., Furchner, J. E. and Trafton, G. H. (1960) Retention of Ba-133 in mice, rats, and dogs. Los Alamos Scientific Laboratory Report LAMS-2455, pp. 24–31.

Richmond, C. R., Furchner, J. E. and Trafton, G. H. (1962a) Retention and excretion of orally administered Ba-133 by mice. Los Alamos Scientific Laboratory Report LAMS-2780, pp. 17–25.

Richmond, C. R., Furchner, J. E. and Trafton, G. H. (1962b) Retention and excretion of orally administered Ba-133 by rats. Los Alamos Scientific Laboratory Report LAMS-2780, pp. 26–34.

Roberts, T. M., Hutchinson, T. C., Paciga, J., Chattopadhyay, A., Jervis, R. E. and Van Loon, J. (1974) Lead contamination around secondary smelters: estimation of dispersal and accumulation by humans. *Science* **186**, 1120–1123.

Rundo, J. (1967) The retention of barium-133 in man. *Int. J. Radiat. Biol.* **13**, 301–302.

Schales, F. (1964) The excretion of thorium X and its daughter products after intravenous injection in man. In: *Assessment of Radioactivity in Man*, Vol. II, pp. 267–276. IAEA, Vienna.

Scheiman-Tagger, E. and Brodie, A. G. (1964) Lead acetate as a marker of growing calcified tissues. *Anat. Rec.* **150**, 435–440.

Schlenker, R. A., Keane, A. T. and Holtzman, R. B. (1982) The retention of Ra-226 in human soft tissue and bone; implications for the ICRP 20 alkaline earth model. *Health Phys.* **42**, 671–693.

Schlundt, H., Nerancy, J. T. and Morris, J. P. (1933) The detection and estimation of radium in living persons. IV. The retention of soluble radium salts administered intravenously. *Am. J. Roentgenol., Radium Ther.* **30**, 515–522.

Schulert, A. R., Peets, E. A., Laszlo, D., Spencer, H., Charles, M. and Samachson, J. (1959) Comparative metabolism of strontium and calcium in man. *Int. J. Appl. Radiat. Isotop.* **4**, 144–153.

Shiels, D. O. (1954) The elimination of lead in sweat. *Austral. Ann. Med.* **3**, 225–229.

Spencer, H., Samachson, M. and Laszlo, D. (1960) Metabolism of strontium-85 and calcium-45 in man. *Metabolism* **9**, 916–925.

Stark, G. (1968) Studies on synthetic hydroxyapatite crystals with regard to metabolism of calcium, strontium, barium and radium in bone. I. The discrimination against calcium. *Biophysik* **5**, 42–54.

Stather, J. W. (1974) Distribution of P-32, Ca-45, Sr-85 and Ba-133 as a function of age in the mouse skeleton. *Health Phys.* **26**, 71–79.

Stather, J. W. (1990) The behaviour, effects and radiation dosimetry of radium in man. In: *IAEA Technical Reports*, Series No. 310, Vol. 2, pp. 297–343. IAEA, Vienna.

Stauber, J. L. and Florence, T. M. (1988) A comparative study of copper, lead, cadmium and zinc in human sweat and blood. *Sci. Total Environ.* **74**, 235–247.

Vander, A. J., Taylor, D. L., Kalitis, K., Mouw, D. R. and Victery, W. (1977) Renal handling of lead in dogs: Clearance studies. *Am. J. Physiol.* **233**(6), F532–F538.

Van Dilla, M. A., Stover, B. J., Floyd, R. L., Atherton, D. R. and Taysum, D. H. (1958) Radium (Ra-226) and radon (Rn-222) metabolism in dogs. *Radiat. Res.* **8**, 417–437.

Verbeeck, R. M. H., Lassuyt, C. J., Heijligers, H. J. M., Driessens, F. C. M. and Vrolijk, J. W. G. A. (1981) Lattice parameters and cation distribution of solid solutions of calcium and lead hydroxyapatite. *Calcif. Tissue Int.* **33**, 243–247.

Victery, W., Vander, A. J. and Mouw, D. R. (1979a) Effect of acid–base status on renal excretion and accumulation of lead in dogs and rats. *Am. J. Physiol.* **237**(5), F398–F407.

Victery, W., Vander, A. J. and Mouw, D. R. (1979b) Renal handling of lead in dogs: stop-flow analysis. *Am. J. Physiol.* **237**(5), F408–F414.

Vincent, J. (1957) Les remaniements de l'os compact marque a l'adie de plomb. *Rev. Belg. Path. Med. Exp.* **26**, 168–172.

Wells, A. C., Venn, J. B. and Heard, M. J. (1975) Deposition in the lung and uptake to blood of motor exhaust labelled with Pb-203. In: *Inhaled Particles IV* (W. H. Walton and B. McGovern, Eds), pp. 175–189. Proceedings of a Symposium of the British Occupational Hygiene Society, Edinburgh.

Wenger, P. and Soucas, K. (1975) Retention and excretion curves of persons containing ^{90}Sr and ^{226}Ra after a chronic contamination. *Health Phys.* **28**, 145–152.

Willes, R. F., Lok, E., Truelove, J. F. and Sundaram, A. (1977) Retention and tissue distribution of Pb-210$(NO_3)_2$ administered orally to infant and adult monkeys. *J. Toxicol. Environ. Health* **3**, 395–406.

Wood, S. K., Farnham, J. E. and Marshall, J. H. (1970) Ca-45, Ba-133, and Ra-226 in 6- to 10-year old beagle dogs: A 100-day study. In: *Radiological Physics Division Annual Report.* Center for Human Radiobiology, July 1969 through June 1970. Argonne National Laboratory ANL-7760, Part II, Biology and Medicine, pp. 110–132. Argonne, IL.

Woodard, H. Q. and Dwyer, A. J. (1972) Whole-body retention of Sr-85 in three children. In: *Second International Conference on Strontium Metabolism*, Glasgow and Strontian, 16–19 August 1972. CONF-720818, pp. 91–109.

Yen, P. K.-J. and Shaw, J. H. (1977) Remodeling of compact bone studied with lead acetate as an intravital stain. *J. Dent. Res.* **56**, 961–966.

APPENDIX B: AGE-SPECIFIC BIOKINETIC MODELS FOR PLUTONIUM, AMERICIUM AND NEPTUNIUM

(B1) The biokinetic models for plutonium, americium, and neptunium described in *ICRP Publication 56* (ICRP, 1989) have been modified for use in the present report. This was done for two reasons:

1. To derive effective doses based on the ICRP's new tissue weighting factors (ICRP, 1991), particularly inclusion of urinary bladder as a listed organ, it is necessary to associate biological removal of activity from the body with specific excretion pathways. This was not done in a uniform manner in the models described in Part 1.
2. New information on the biokinetics of plutonium and americium in humans has been derived since the completion of the models given in Part 1. Although this information is consistent in most respects with the earlier data and with the models used in Part 1, the newer data suggest greater retention of the transuranic elements in some soft tissues than was previously predicted, particularly at times remote from exposure.

(B2) The model structure (Fig. B1) adopted here for application to plutonium, americium and neptunium is the same as that proposed in a recently published age-specific biokinetic model for americium (Leggett, 1992), except that a second liver compartment has been included for consideration of plutonium and neptunium. The present models for plutonium, americium and neptunium are not substantially different from those in Part 1 of this document with regard to uptake and retention by the skeleton, liver and gonads, but uptake and retention of these elements by other soft tissues and their removal from the body in excreta are now treated in more detail.

(B3) As in Part 1, the skeleton is divided into cortical and trabecular fractions, and each of these is subdivided into bone surfaces, bone volume and a bone cavity compartment containing soft tissue. For purposes of dosimetry, the bone cavity compartment is assumed to consist entirely of bone marrow. Activity entering the skeleton is assigned initially to bone surfaces and is subsequently transferred to bone marrow by bone resorption or to bone volume by bone formation. Activity in bone volume is transferred to bone marrow by bone resorption. Activity is removed from bone marrow to blood over a period of months and is redistributed in the same pattern as the original input to blood.

(B4) Blood is treated as a uniformly mixed pool. Compartment ST0 is a soft tissue pool that includes the extracellular fluids and exchanges material with blood over a period of hours or days. It is used in the model for two purposes: to depict an early build-up and decline of material in soft tissues, and to account for early feedback of material to blood. Thus, compartment ST0 represents an integral part of the early circulation of material. In the summary of parameter values below, deposition fractions for compartments other than ST0 are given in terms of "activity leaving the circulation" and refer to the division of material among compartments other than ST0.

(B5) Soft tissue compartments ST1 and ST2 are used to represent intermediate-term retention (up to 2 years) and tenacious retention (many years), respectively, in the "massive soft tissues" (muscle, skin, subcutaneous fat and all other soft tissues not explicitly included in other compartments of the model).

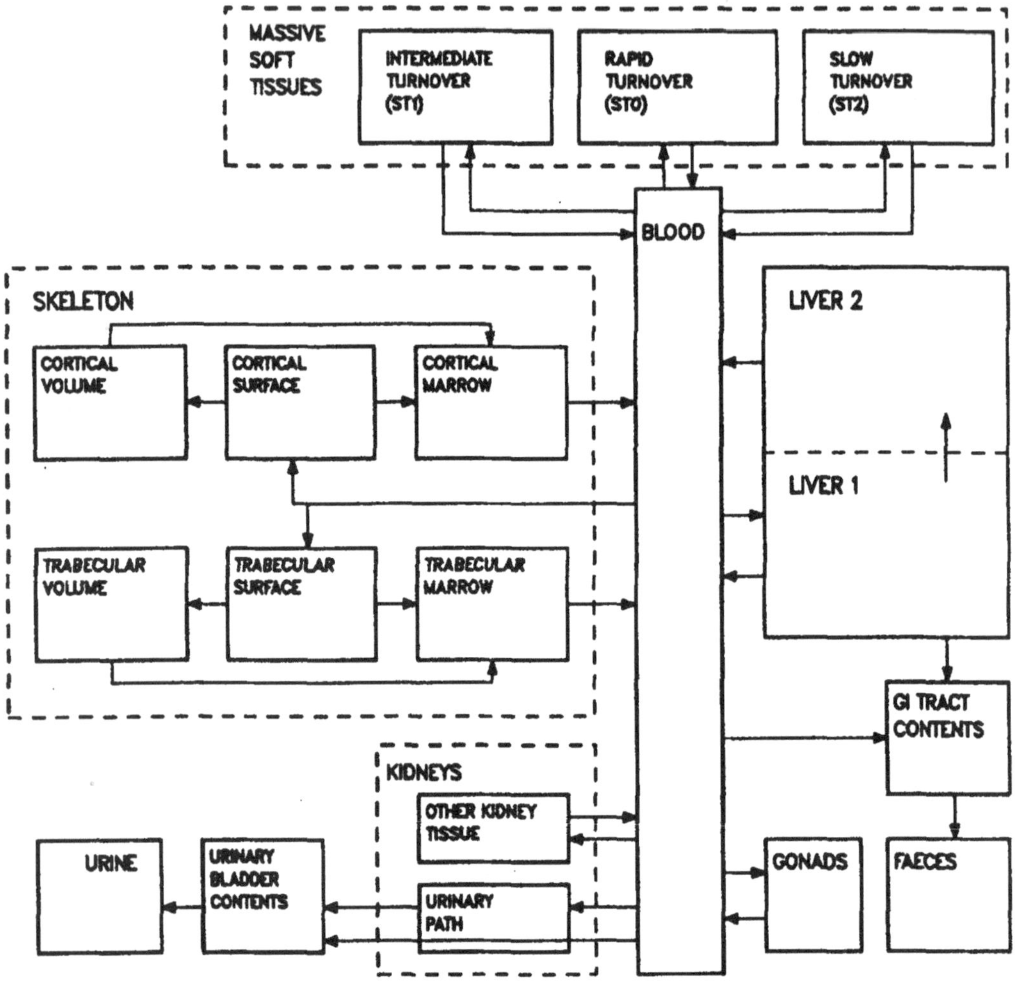

Fig. B1. Diagram of the biokinetic model for plutonium, americium and neptunium.

(B6) Depending on the element, the liver is viewed as a single compartment, indicated as Liver 1, or two compartments, with the second compartment indicated as Liver 2. Liver 2 represents relatively tenacious retention ($T_{1/2} > 1$ year) in the liver; it is defined on a kinetic rather than a biological basis but appears to be associated largely with the reticuloendothelial (RE) system. Data on hepatic retention and faecal excretion of plutonium are most easily reproduced using a two-compartment liver model, with one compartment losing a portion of its activity to the gastrointestinal tract over a relatively short period (1 year) and the second showing greater retention of plutonium (many years). On the other hand, data on americium, which may be taken up and/or retained to a lesser extent than plutonium in RE cells, can be reproduced reasonably well if the liver is treated as a uniformly mixed pool that loses activity to blood and the gastrointestinal tract with a biological half-time of 1 year. Material depositing in the liver enters Liver 1, from which part of the material is removed to the gastrointestinal tract contents via biliary secretion, and the rest is removed either to blood (americium) or to Liver 2 (plutonium, neptunium). Material leaving Liver 2 is assigned to blood.

(B7) The kidneys are assumed to consist of two compartments, one that loses activity to urine and another that returns activity to blood. The "urinary bladder contents" is considered as a separate pool that receives all material destined for urinary excretion.

(B8) Movement of material in the body is depicted as a system of first-order processes. Parameter values are expressed as transfer rates (that is, fractional flows per unit time) between compartments, but most of the derived transfer rates are secondary values calculated from selected "deposition fractions" and "removal half-times". The concepts of "deposition fraction" and "removal half-time" are commonly used in connection with non-recycling biokinetic models but require some explanation with regard to the recycling models used in this report.

(B9) The "removal half-time" from a compartment refers to the biological half-time that would be observed if there were no recycling to that compartment. This will generally differ from the apparent (or net, or externally viewed) half-time that may be estimated at any given time in the presence of recycling. For example, the removal half-time of americium from compartment ST0 to blood is assumed to be 0.5 days (that is, the transfer rate from ST0 to blood is $\ln(2)/0.5\ \text{day}^{-1} = 1.386\ \text{day}^{-1}$), but it would require more than 0.5 days for the contents of ST0 to be reduced by 50% because of a continual feed from blood. Similarly, the removal half-time of americium from blood (by all pathways combined) is assumed to be 30 min, corresponding to a transfer rate to all destinations of $\ln(2)/0.5\ \text{day}^{-1} = 33.27\ \text{day}^{-1}$, but the disappearance curve from blood at 1 day suggests a much longer half-time because flow to blood from ST0 and other compartments would largely offset outflow from blood at that time.

(B10) Transfer rates from blood to various compartments are based on liberally rounded "deposition fractions", which provide a convenient way to describe the initial distribution of activity leaving the circulation. For example, the transfer rate of americium from blood to Liver 1 is based on the estimates that 30% of activity leaving blood is deposited in the rapid-turnover soft-tissue compartment ST0, and 50% of americium leaving the circulation (i.e. leaving blood but not depositing in ST0) deposits in Liver 1. Since the total removal rate from blood is $33.27\ \text{day}^{-1}$ and 30% of this goes to ST0, the transfer rate to all other destinations is $0.7 \cdot 33.27\ \text{day}^{-1} = 23.29\ \text{day}^{-1}$. Therefore, the transfer rate of americium from blood to Liver 1 is $0.5 \cdot 23.29\ \text{day}^{-1} = 11.6\ \text{day}^{-1}$.

(B11) Parameter values for americium, plutonium, and neptunium are given in Tables B-1, B-2, and B-3, respectively, and are summarized below.

Uptake to blood

(B12) *Adults.* In *ICRP Publication 48* (ICRP, 1989) the gastrointestinal absorption of plutonium (Pu), americium (Am) and neptunium (Np) was extensively reviewed. It was concluded that an f_1 of 10^{-3} would provide an adequate margin of safety for ingestion of unknown forms or mixed compounds of plutonium by adult humans. This same value was recommended for adults by an Expert Group of the NEA/OECD (1988) and was used in Part 1 (ICRP, 1989). The recommendation was based very largely on results from animal experiments supported by limited human data. A study by Maletskos *et al.* (1969) of the absorption of ^{234}Th ingested as the sulphate in a "mock dial paint" by six 63 to 83-year-old humans gave f_1 values in the range 10^{-4} to

Table B-1. Age-specific transfer rates (d^{-1}) for americium model.

	Age					
	3 mo	1 y	5 y	10 y	15 y	Adult
blood to Liver 1	2.33	2.33	6.98	6.98	6.98	11.6
blood to cortical surface	8.15	8.15	5.82	5.82	5.82	3.49
blood to trabecular surface	8.15	8.15	5.82	5.82	5.82	3.49
blood to urinary bladder content	1.63	1.63	1.63	1.63	1.63	1.63
blood to kidney (urinary path)	0.466	0.466	0.466	0.466	0.466	0.466
blood to other kidney tissue	0.116	0.116	0.116	0.116	0.116	0.116
blood to ULI contents	0.303	0.303	0.303	0.303	0.303	0.303
blood to testes	0.00047	0.00070	0.00079	0.00093	0.0075	0.0082
blood to ovaries	0.00028	0.00037	0.00093	0.0016	0.0028	0.0026
blood to ST0	10.0	10.0	10.0	10.0	10.0	10.0
blood to ST1	1.67	1.67	1.67	1.67	1.67	1.67
blood to ST2	0.466	0.466	0.466	0.466	0.466	0.466
ST0 to blood	1.386	1.386	1.386	1.386	1.386	1.386
kidneys (urinary path) to bladder	0.099	0.099	0.099	0.099	0.099	0.099
other kidney tissue to blood	0.00139	0.00139	0.00139	0.00139	0.00139	0.00139
ST1 to blood	0.0139	0.0139	0.0139	0.0139	0.0139	0.0139
ST2 to blood	0.000019	0.000019	0.000019	0.000019	0.000019	0.000019
trabecular surface to volume	0.00822	0.00288	0.00181	0.00132	0.000959	0.000247
trabecular surface to marrow	0.00822	0.00288	0.00181	0.00132	0.000959	0.000493
cortical surface to volume	0.00822	0.00288	0.00153	0.000904	0.000521	0.0000411
cortical surface to marrow	0.00822	0.00288	0.00153	0.000904	0.000521	0.0000821
trabecular volume to marrow	0.00822	0.00288	0.00181	0.00132	0.000959	0.000493
cortical volume to marrow	0.00822	0.00288	0.00153	0.000904	0.000521	0.0000821
cort/trab bone marrow to blood	0.0076	0.0076	0.0076	0.0076	0.0076	0.0076
Liver 1 to blood	0.00185	0.00185	0.00185	0.00185	0.00185	0.00185
Liver 1 to small intestine	0.000049	0.000049	0.000049	0.000049	0.000049	0.000049
gonads to blood	0.00019	0.00019	0.00019	0.00019	0.00019	0.00019
f_1	0.005	0.0005	0.0005	0.0005	0.0005	0.0005

$6 \cdot 10^{-4}$, with a mean of $2 \cdot 10^{-4}$. This study is of interest since thorium can, with caution, be regarded as an analogue for plutonium. The absorption of fallout plutonium from reindeer meat was measured by Mussalo-Rauhamaa *et al.* (1984). An f_1 value of $8 \cdot 10^{-4}$ was obtained by comparing the ratio of body content to dietary intake of $^{239/240}Pu$ in persons who had lived in Lappland or the urban areas of southern Finland. Large uncertainties were associated with this estimate of absorption. A preliminary study of the absorption of ^{239}Np in human volunteers, following its absorption as the citrate, indicated an f_1 value of about $2 \cdot 10^{-4}$ (Popplewell *et al.*, 1989). These values

Table B-2. Age-specific transfer rates (d^{-1}) for plutonium model.

	Age					
	3 mo	1 y	5 y	10 y	15 y	Adult
blood to Liver 1	0.0647	0.0647	0.1294	0.1294	0.1294	0.1941
blood to cortical surface	0.2264	0.2264	0.1941	0.1941	0.1941	0.1294
blood to trabecular surface	0.2264	0.2264	0.1941	0.1941	0.1941	0.1941
blood to urinary bladder content	0.0129	0.0129	0.0129	0.0129	0.0129	0.0129
blood to kidney (urinary path)	0.00647	0.00647	0.00647	0.00647	0.00647	0.00647
blood to other kidney tissue	0.00323	0.00323	0.00323	0.00323	0.00323	0.00323
blood to ULI contents	0.0129	0.0129	0.0129	0.0129	0.0129	0.0129
blood to testes	0.000013	0.000019	0.000022	0.000026	0.00021	0.00023
blood to ovaries	0.000008	0.000010	0.000026	0.000045	0.000078	0.000071
blood to ST0	0.2773	0.2773	0.2773	0.2773	0.2773	0.2773
blood to ST1	0.0806	0.0806	0.0806	0.0806	0.0806	0.0806
blood to ST2	0.0129	0.0129	0.0129	0.0129	0.0129	0.0129
ST0 to blood	0.693	0.693	0.693	0.693	0.693	0.693
kidneys (urinary path) to bladder	0.01386	0.01386	0.01386	0.01386	0.01386	0.01386
other kidney tissue to blood	0.00139	0.00139	0.00139	0.00139	0.00139	0.00139
ST1 to blood	0.000475	0.000475	0.000475	0.000475	0.000475	0.000475
ST1 to urinary bladder contents	0.000475	0.000475	0.000475	0.000475	0.000475	0.000475
ST2 to blood	0.000019	0.000019	0.000019	0.000019	0.000019	0.000019
trabecular surface to volume	0.00822	0.00288	0.00181	0.00132	0.000959	0.000247
trabecular surface to marrow	0.00822	0.00288	0.00181	0.00132	0.000959	0.000493
cortical surface to volume	0.00822	0.00288	0.00153	0.000904	0.000521	0.0000411
cortical surface to marrow	0.00822	0.00288	0.00153	0.000904	0.000521	0.0000821
trabecular volume to marrow	0.00822	0.00288	0.00181	0.00132	0.000959	0.000493
cortical volume to marrow	0.00822	0.00288	0.00153	0.000904	0.000521	0.0000821
cort/trab bone marrow to blood	0.0076	0.0076	0.0076	0.0076	0.0076	0.0076
Liver 1 to Liver 2	0.00177	0.00177	0.00177	0.00177	0.00177	0.00177
Liver 1 to small intestine	0.000133	0.000133	0.000133	0.000133	0.000133	0.000133
Liver 2 to blood	0.000211	0.000211	0.000211	0.000211	0.000211	0.000211
gonads to blood	0.00019	0.00019	0.00019	0.00019	0.00019	0.00019
f_1	0.005	0.0005	0.0005	0.0005	0.0005	0.0005

are valuable in providing human data to support the assumption that absorption in man will be within the range of absorption in experimental animals.

(B13) Two recent studies have provided further data on the absorption of actinides in humans. These are absorption measurements of plutonium and americium administered in citrate solution (Popplewell *et al.*, 1991). Combining the results of two separate studies, in each of which the urinary excretion of $^{239/240}Pu$ and ^{241}Am was

Table B-3. Age-specific transfer rates (d^{-1}) for neptunium model.

	Age					
	3 mo	1 y	5 y	10 y	15 y	Adult
blood to Liver 1	0.0582	0.0582	0.0970	0.0970	0.0970	0.1941
blood to cortical surface	0.8831	0.8831	0.6405	0.6405	0.6405	0.3930
blood to trabecular surface	0.4755	0.4755	0.5240	0.5240	0.5240	0.4804
blood to urinary bladder content	0.2717	0.2717	0.4270	0.4270	0.4270	0.6211
blood to kidney (urinary path)	0.0291	0.0291	0.0291	0.0291	0.0291	0.0291
blood to other kidney tissue	0.0097	0.0097	0.0097	0.0097	0.0097	0.0097
blood to ULI contents	0.01359	0.01359	0.01359	0.01359	0.01359	0.01359
blood to testes	0.000039	0.000058	0.000066	0.000078	0.00062	0.00068
blood to ovaries	0.000023	0.000031	0.000078	0.00014	0.00023	0.00021
blood to ST0	0.832	0.832	0.832	0.832	0.832	0.832
blood to ST1	0.1611	0.1611	0.1611	0.1611	0.1611	0.1611
blood to ST2	0.0388	0.0388	0.0388	0.0388	0.0388	0.0388
ST0 to blood	0.6931	0.6931	0.6931	0.6931	0.6931	0.6931
kidneys (urinary path) to bladder	0.0495	0.0495	0.0495	0.0495	0.0495	0.0495
other kidney tissue to blood	0.00139	0.00139	0.00139	0.00139	0.00139	0.00139
ST1 to blood	0.00693	0.00693	0.00693	0.00693	0.00693	0.00693
ST2 to blood	0.000019	0.000019	0.000019	0.000019	0.000019	0.000019
trabecular surface to volume	0.00822	0.00288	0.00181	0.00132	0.000959	0.000247
trabecular surface to marrow	0.00822	0.00288	0.00181	0.00132	0.000959	0.000493
cortical surface to volume	0.00822	0.00288	0.00153	0.000904	0.000521	0.0000411
cortical surface to marrow	0.00822	0.00288	0.00153	0.000904	0.000521	0.0000821
trabecular volume to marrow	0.00822	0.00288	0.00181	0.00132	0.000959	0.000493
cortical volume to marrow	0.00822	0.00288	0.00153	0.000904	0.000521	0.0000821
cort/trab bone marrow to blood	0.0076	0.0076	0.0076	0.0076	0.0076	0.0076
liver 1 to liver 2	0.00177	0.00177	0.00177	0.00177	0.00177	0.00177
liver 1 to small intestine	0.000133	0.000133	0.000133	0.000133	0.000133	0.000133
liver 2 to blood	0.0019	0.0019	0.0019	0.0019	0.0019	0.0019
gonads to blood	0.00019	0.00019	0.00019	0.00019	0.00019	0.00019
f_1	0.005	0.0005	0.0005	0.0005	0.0005	0.0005

measured in six men and two women after consumption of winkles (Hunt *et al.*, 1986, 1989, 1990), the estimated absorption values were $2 \cdot 10^{-4}$ (range $2 \cdot 10^{-5}$ to $4.9 \cdot 10^{-4}$) for plutonium and $1 \cdot 10^{-4}$ (range $4 \cdot 10^{-5}$ to $2.8 \cdot 10^{-4}$) for americium. The results for ^{239}Np and ^{242}Cm absorption in five volunteers, obtained by comparing urinary excretion after intravenous and oral administration of the nuclides as their citrate complexes, were $2 \cdot 10^{-4}$ with range of $1 \cdot 10^{-4}$ to $3 \cdot 10^{-4}$ for each element (Popplewell *et al.*, 1991). On the basis of these results showing similar levels of absorption for these four actinide elements, and taking account of animal data showing variations in f_1 values resulting from differences in chemical form, it is now considered that an appropriate general f_1 value for unknown forms of plutonium, americium and neptunium, including uptake from food, is $5 \cdot 10^{-4}$. This value is adopted in this report.

(B14) *Infants and children.* As described in paragraph (11) of the Introduction to this report, the absorption of radionuclides tends to be greater in the newborn, although the results of animal studies suggest enhancement of gut transfer progressively decreases with increasing age, reaching adult values by about the time of weaning in most cases. For most elements f_1 values for adults can therefore be considered to apply to children of 1 year of age or over. Following the recommendations of an NEA Expert Group (NEA/OECD, 1988), where no human or animal data are available, a general approach is adopted in this report. For f_1 values between 0.01 and 0.5 in the adult, an increase by a factor of 2 is assumed for the first year of life, but for elements with an f_1 in the adult of 0.001 or less a value of 10 times the adult value is assumed. This general approach was adopted in Part 1 (ICRP, 1989). On this basis an f_1 value of $5 \cdot 10^{-3}$ for plutonium, americium and neptunium is adopted in this report for 3-month-old infants. For children of 1 year and older an f_1 of $5 \cdot 10^{-4}$ is used.

Summary of Parameter Values for Americium

(B15) Parameter values for americium are taken directly from the americium model of Leggett (1992), except for the following modifications:

1. The model of Leggett addresses changes throughout life in the rate of bone turnover. In the present model, now adopted, bone turnover rates are assumed to remain constant in the adult. The assumed bone turnover rates for cortical and trabecular bone for different age groups are the same as those used in Part 1 and are equal to the transfer rates from cortical and trabecular bone volume, respectively, to bone marrow, given in Table B-1.
2. Parameter values describing uptake and retention by gonadal tissue of the adult are changed for consistency with earlier recommendations of the ICRP (ICRP, 1986, 1989). Also, assumptions regarding uptake and retention by gonadal tissues of children are more cautious with regard to estimates of dose equivalent than those used in the model of Leggett.

Removal from the circulation

(B16) There is little information on the behaviour of americium in the human circulation. Data on short-term retention of americium in blood and massive soft tissues of laboratory animals (McClellan *et al.*, 1962; Turner and Taylor, 1968; Belyaev, 1969; Bruenger *et al.*,1969; Atherton and Lloyd, 1972; Rosen *et al.*, 1972; Stevens and Bruenger, 1972; Cohen and Wrenn, 1973; Durbin, 1973; Sutton *et al.*, 1978; Guilmette *et al.*, 1980) are represented reasonably well by treating blood as a uniformly mixed pool that loses activity at a rate of 33.3 day^{-1} (corresponding to a removal half-time of 30 min) and postulating a soft tissue compartment, ST0, that receives 30% of americium leaving blood and returns americium to blood with a relatively short half-time, taken as 0.5 days. Parameter values describing removal from the circulation are independent of age.

Deposition in the liver and skeleton

(B17) The initial distribution of americium leaving the circulation is estimated from americium injection data for beagles (Lloyd *et al.*, 1970, 1972, 1984), monkeys (Durbin, 1973), and baboons (Guilmette *et al.*, 1980) and external measurements of humans accidentally exposed to americium (Fry, 1976; Cohen *et al.*, 1979; Toohey and Essling, 1980; Robinson *et al.*, 1983; Popplewell and Ham, 1989; Doerfel and Oliveira, 1989). For all ages at injection, liver and bone surfaces combined are assumed to receive 80% of americium leaving the circulation, but the division of this amount between liver and bone surfaces varies with age. Relative deposition in (a) skeleton and (b) liver are set at a:b = 3:5 for adults, based on data for humans, dogs, monkeys, and baboons for early times after exposure. Ratios a:b of 7:1 and 5:3 are assumed for the first year of life (through age 1 year) and ages 5 to 15 years, respectively, based mainly on analogy with data for laboratory animals injected with americium and plutonium during various stages of life. For all ages, half of the skeletal deposition is assigned to cortical bone surfaces and half is assigned to trabecular surfaces.

Translocation of skeletal deposits

(B18) The model for translocation of skeletal deposits is the same as that for plutonium, americium and neptunium given in Part 1 (ICRP, 1989). According to that model, the rate of translocation of skeletal deposits of americium is controlled by bone restructuring processes. The age-specific transfer rate from compact or trabecular bone surfaces to the corresponding bone marrow compartment is equal to the rate at which that type of bone surface is resorbed. The age-specific transfer rate from bone surfaces to bone volume is equal to the surface formation rate in persons of age 0 to 15 years and one-half the surface formation rate in adults. The age-specific transfer rate from compact or trabecular bone volume to the corresponding bone marrow compartment is equal to the rate at which that type of bone volume is resorbed. The age-specific bone formation rates are the same as those used in Part 1 and, as in that document, a common value is used for bone formation and resorption and is applied both to surface and volume remodelling. The removal half-time from bone cavities is assumed to be 0.25 years (rounded from the value of 90 days used in Part 1). This value was originally derived for plutonium and was based on data for beagles. Model predictions based on this removal rate are generally consistent with recent measurements of plutonium and americium in bone and bone marrow of occupationally exposed persons (Kathren, 1989, 1990; McInroy and Kathren, 1990). All material leaving bone cavities is assumed to enter blood. Figure B2 gives estimated americium contents of skeleton as a function of time after injection of americium into infants, 10-year-old children, or mature adults, based on the parameter values given in Table B-1.

Removal from the liver

(B19) At all ages, americium is assumed to be removed from the liver (compartment Liver 1) with a half-time of 1 year, based on data for occupationally exposed humans (Durbin and Schmidt, 1985; Kathren *et al.*, 1988; McInroy *et al.*, 1989; Popplewell and Ham, 1989) and to a smaller extent on data for americium-injected beagles (as reanalyzed by Griffith *et al.* (1983) and Leggett (1992)). The model predicts an "externally viewed half-time" in liver of 2 to 3 years in the first few years after injection of americium to blood; at times remote from injection (decades) there would be little or no apparent loss from liver because, as the liver content falls to a few per cent of the total systemic burden, the amount fed back to the liver from blood each day would be about the same as the amount leaving the liver. It is assumed that americium leaving Liver 1 returns to blood, except for a small quantity moving via biliary secretion to the contents of the gastrointestinal tract. Figure B3 gives estimated americium contents of liver as a function of time after injection of americium into infants, 10-year-old children, or mature adults, based on the parameter values given in Table B-1.

Uptake and removal by soft tissue compartments ST1 and ST2

(B20) Of activity leaving the circulation, 2% is assigned to the soft tissue compartment ST2 and the percentage left over after all other deposition fractions in the model have been chosen, slightly more than 7%, is assigned to the soft tissue compartment ST1. The removal half-times from compartments ST1 and ST2 are set at 50 days and 100 years, respectively. These deposition fractions and removal half-times yield reasonable agreement with measurements of americium in massive soft tissues of beagles (Lloyd *et al.*, 1970; Atherton and Lloyd, 1972), monkeys (Durbin, 1973) and

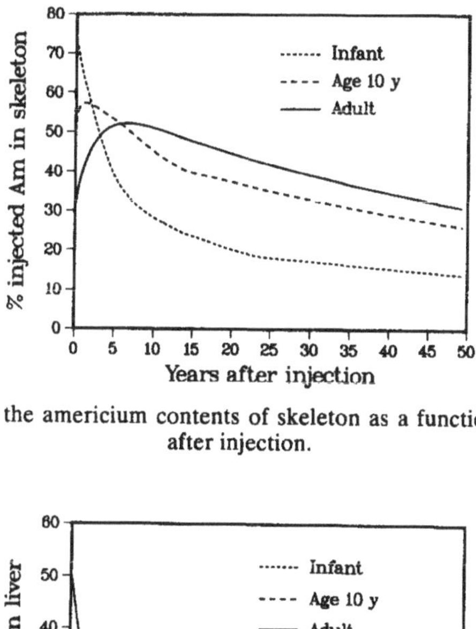

Fig. B2. Model predictions of the americium contents of skeleton as a function of age at injection and time after injection.

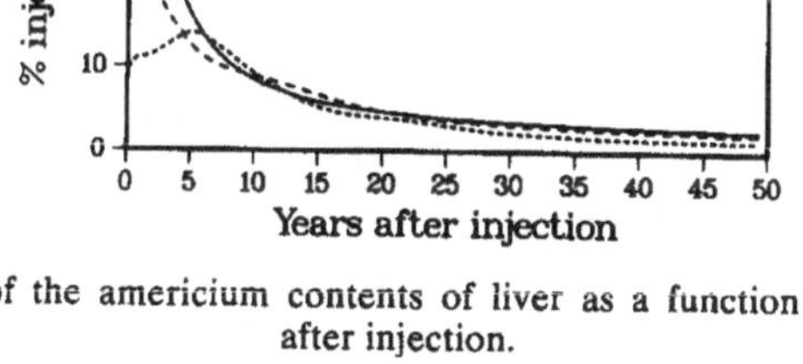

Fig. B3. Model predictions of the americium contents of liver as a function of age at injection and time after injection.

baboons (Guilmette *et al.*, 1980) at various times after injection and in massive soft tissues of humans many years after occupational exposure to americium (McInroy *et al.*, 1989). Parameter values describing uptake and retention in ST1 and ST2 are independent of age.

Uptake and removal by the gonads

(B21) The model for uptake and removal of americium by gonads in the adult is based on recommendations given in *ICRP Publication 48* (ICRP, 1986), but the retention time is adjusted to consider recycling of material. For the adult it is assumed that deposition in the gonads, expressed as a percentage of americium leaving the circulation, is 0.001% g^{-1} gonadal tissue. This yields a deposition of 0.035% of americium leaving the circulation in the 35 g testes of the reference adult male and 0.011% in the 11 g ovaries of the reference adult female (ICRP, 1975). For non-adults the assumption is made that fractional uptake per gram of gonadal tissue is twice that in adults, or 0.002% g^{-1} gonadal tissue; this is intended as a cautious assumption with regard to estimated dose equivalent to gonadal tissues, as judged from comparative gonadal uptake of plutonium in immature beagles and in mature beagles (Miller *et al.*, 1989). Based on data given in the *ICRP Publication 23* (ICRP, 1975), the weight of the testes is assumed to be 1 g in infants, 1.5 g at age 1 year, 1.7 g at age 5 years, 2 g at age 10 years, and 16 g at age 15 years; the weight of the ovaries is assumed to be 0.6 g in infants, 0.8 g at age 1 year, 2.0 g at age 5 years, 3.5 g at age 10 years, and 6 g at age 15 years. As in Part 1 (ICRP, 1989), the removal half-time from gonads to blood is assumed to be 10 years for all age groups. This value is reasonably consistent with the assumption in *ICRP Publication 48* (ICRP, 1986) of permanent retention in gonadal tissue due to consideration in the present model of continuous gonadal uptake of activity lost from other tissues.

The model for urinary excretion and for retention in the kidneys

(B22) Of activity leaving the circulation, 7% is assumed to move directly to the urinary bladder, and another 2% is assumed to deposit in the urinary path in the kidneys and to be released to the urinary bladder contents with a half-time of 7 days. Also 0.5% of activity leaving the circulation is assumed to deposit in other kidney tissues and to be returned to blood with a half-time of 500 days. These values were based on urinary excretion rates observed in the first 1–3 weeks in beagles (Lloyd *et al.*, 1970), monkeys (Durbin, 1973) and baboons (Rosen *et al.*, 1972; Cohen and Wrenn, 1973) injected with americium and in humans (Popplewell *et al.*, 1989, 1991) and beagles (Lloyd *et al.*, 1974) injected with curium; collective animal data on the amount and microlocalization of americium deposited in the kidneys as a function of time after injection (Scott *et al.*, 1948; Zalikin *et al.*, 1968; Lloyd *et al.*, 1970; McKay *et al.*, 1972; Atherton and Lloyd, 1972; Wright, 1972; Durbin, 1973; Moskalev, 1977; Guilmette *et al.*, 1980); urinary excretion data on humans accidentally exposed to americium (Whalen and Davies, 1972; Wrenn *et al.*, 1972; King, 1980; Rosen *et al.*, 1980; Robinson *et al.*, 1983; Takada *et al.*, 1984; Heid and Robinson, 1985; Breitenstein and Palmer, 1989), and autopsy measurements of americium in kidneys of occupationally exposed persons at times remote from exposure (McInroy *et al.*, 1989). Parameter values describing urinary excretion and retention in the kidneys are independent of age.

The model for faecal excretion

(B23) Faecal excretion of americium is assumed to arise from two sources: secretion from blood into the gastrointestinal tract, amounting to 1.3% of material leaving the circulation, and biliary secretion from the liver into the gastrointestinal tract, amounting to 2.6% of material leaving the liver. Transfer rates from blood and Liver 1 (the latter being the assigned source of biliary secretion) to the gastrointestinal tract were based on consideration of observed ratios of urinary to faecal excretion in human studies (Robinson *et al.*, 1983; Heid and Robinson, 1985; Breitenstein and Palmer, 1989; Durbin and Schmidt, 1985) and the assumption that, in the adult human, half of faecal excretion is derived from biliary secretion and half from other intestinal secretions. Parameter values describing faecal excretion are independent of age.

Summary of Parameter Values for Plutonium

Removal from the circulation

(B24) Data on early kinetics of plutonium in the human circulation (Langham *et al.*, 1950; Durbin, 1972; Talbot *et al.*, 1992) can be represented reasonably well by treating blood as a uniformly mixed pool that loses activity with a half-time of 0.75 days and postulating a soft tissue compartment (ST0) in relatively rapid exchange with blood. By analogy with americium, the soft tissue compartment ST0 is assumed to receive 30% of plutonium leaving blood. The removal half-time from ST0 is assumed to be 1 day, based on soft tissue data for humans and non-human primates injected with plutonium (Langham *et al.*, 1950; Durbin, 1972; Durbin *et al.*, 1985). Parameter values describing removal from the circulation are independent of age.

Deposition in the liver and skeleton

(B25) Assumptions concerning fractional deposition of plutonium in the liver (Liver 1) and on trabecular and cortical bone surfaces are essentially the same as in Part 1 (ICRP, 1989), except that (as for americium) the fractional deposition assigned to infants (age 3 months) is also assumed for age 1 year in this document. For all ages at injection, liver and bone surfaces combined are assumed to receive 80% of plutonium leaving the circulation, but division of this amount between liver and bone surfaces varies with age. Relative deposition in (a) skeleton and (b) liver are set at a:b = 5:3 for adults. As in the model for americium, a ratio a:b of 7:1 is assigned to the infant and 1-year-old child. A ratio intermediate to those for the 1-year-old child and adult, a:b = 6:2, is applied to ages 5 to 15 years. For adults it is assumed that trabecular surfaces receive 60% and cortical surfaces 40% of the amount depositing in the skeleton (ICRP, 1989). For pre-adults it is assumed that half of the skeletal deposition is on cortical bone surfaces and half is on trabecular surfaces (ICRP, 1989).

Translocation of skeletal deposits

(B26) The model for translocation of skeletal deposits of plutonium is the same as that for plutonium, americium, and neptunium given in Part 1 (ICRP, 1989) and is summarized above in the discussion of parameter values for americium. Figure B4 gives estimated plutonium contents of skeleton as a function of time after injection of plutonium into infants, 10-year-old children, or mature adults, based on the parameter values given in Table B-2.

Removal from the liver

(B27) The model for retention in the liver is effectively the same as that described in Part 1 with regard to integrated activity, but a second liver compartment is used here to describe biliary excretion of plutonium, which was not considered explicitly in Part 1. It is assumed that plutonium depositing in the liver enters a first compartment, indicated as Liver 1. The removal half-time from Liver 1 is assumed to be the same as that for americium (1 year); however, plutonium leaving Liver 1 is assumed to flow to a second liver compartment, Liver 2, except for a small amount assigned to biliary excretion, as discussed later. Plutonium in Liver 2 is assumed to be removed to blood with a half-time of 9 years, yielding an effective removal half-time from the total liver of about 10 years, the value used in Part 1 (ICRP, 1989). This model of the liver, together with the assumptions made here for the skeleton and other soft tissues, yields predictions of the long-term distribution of plutonium between liver, skeleton and other soft tissues that are consistent with autopsy data on occupationally exposed subjects (Kathren *et al.*, 1988; McInroy *et al.*, 1989). By contrast, the retention model for plutonium implied in *ICRP Publication 48* substantially underestimates the portion of total-body plutonium found in the liver in these subjects (McInroy *et al.*, 1989). Figure B5 gives estimated plutonium contents of liver as a function of time after injection of plutonium into

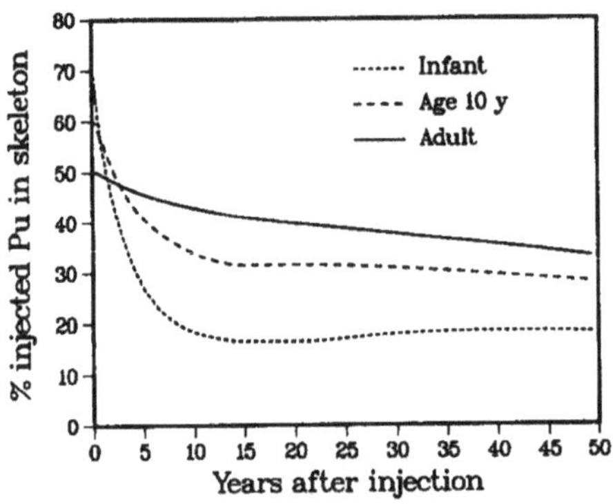

Fig. B4. Model predictions of the plutonium contents of skeleton as a function of age at injection and time after injection.

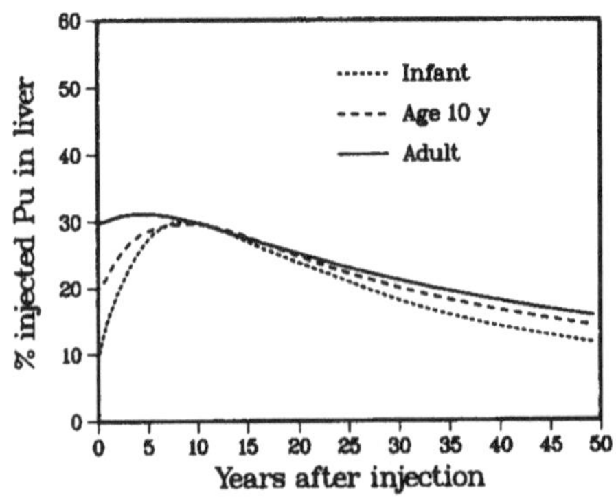

Fig. B5. Model predictions of the plutonium contents of liver as a function of age at injection and time after injection.

infants, 10-year-old children, or mature adults, based on parameter values given in Table B-2.

Uptake and removal by soft tissue compartments ST1 and ST2

(B28) As in the model for americium, 2% of activity leaving the circulation is assigned to the slow-turnover soft tissue compartment ST2, and the percentage left over after all other deposition fractions in the model have been chosen, amounting to almost 12.5% of plutonium leaving the circulation, is assigned to the intermediate-turnover soft tissue compartment ST1. The removal half-time from ST1 is assumed to be 2 years, based on autopsy measurements on human subjects injected with plutonium at times up to 456 days before their deaths (Langham *et al.*, 1950; Durbin, 1972). The removal half-time from ST2 is assumed to be 100 years, the same as for americium; this value yields broad agreement with autopsy data on United States Transuranium Registry (USTR) subjects whose primary exposures to plutonium occurred many years before their deaths (McInroy *et al.*, 1989). Parameter values describing uptake and retention in ST1 and ST2 are independent of age.

Uptake and removal by the gonads

(B29) Assumptions concerning gonadal uptake and retention are the same as those described earlier (B21) for americium.

The model for urinary excretion and for retention in the kidneys

(B30) The model for urinary excretion is designed to yield reasonable agreement with data on human subjects injected with plutonium for experimental purposes or occupationally exposed to this element (Langham *et al.*, 1950; Durbin, 1972; Rundo *et al.*, 1976; Voelz *et al.*, 1979, 1985; Rundo, 1979; Moss *et al.*, 1983; Jones, 1985; Leggett, 1985; Leggett and Eckerman, 1987; Kathren and McInroy, 1991; Talbot *et al.*, 1992). It is assumed that 2% of activity leaving the circulation goes directly to the urinary bladder, that another 1% deposits in the urinary path in the kidneys and is released to the urinary bladder contents with a half-time of 50 days, and that half of the plutonium released from soft tissue compartment ST1 goes directly to the urinary bladder (not shown in Fig. B1). The assumption of direct movement of plutonium from ST1 to the urinary bladder is used to account for an apparent increase with time in fractional clearance of circulating plutonium, as judged from collective data of Langham *et al.* (1950), Durbin (1972), Voelz *et al.* (1979) and Talbot *et al.* (1992). Such an increase in urinary clearance might result, for example, if part or all of the systemically deposited plutonium is released to blood in a chemical form different from that of the original input to blood, although it is not known whether this actually occurs. Finally, as in the model for americium, 0.5% of activity leaving the circulation is assigned to other kidney tissues, from which it is returned to blood with a half-time of 500 days; this yields broad agreement with data on laboratory animals and humans exposed to plutonium (Langham *et al.*, 1950; Durbin, 1972, 1973; Lloyd *et al.*, 1976; 1978a,b; Durbin *et al.*, 1985; McInroy *et al.*, 1989). Parameter values describing urinary excretion and retention in the kidneys are independent of age.

The model for faecal excretion

(B31) It is assumed that 2% of plutonium leaving the circulation and 7% of that leaving Liver 1 goes to the gastrointestinal tract contents. Fractional flow from blood to the gastrointestinal tract was set first to approximate the early faecal excretion rate observed in human subjects (Langham *et al.*, 1950; Durbin, 1972; Talbot *et al.*, 1992), and the rate of biliary excretion was then set to approximate observed rates of faecal excretion and urinary-to-faecal ratios at later times (Langham *et al.*, 1950; Durbin, 1972; Rundo *et al.*, 1976; Rundo, 1979; Voelz *et al.*, 1979). As in the model for americium, the contribution from biliary secretion represents about half of the total amount going to faeces in the adult at times remote from exposure. Parameter values describing faecal excretion are independent of age.

Summary of Parameter Values for Neptunium

(B32) Experimental data on neptunium were summarized and referenced in Part 1 (ICRP, 1989). Where direct information on the removal of neptunium from soft tissue compartments is lacking, it is generally assumed here that removal half-times increase in the order Am ≤ Np ≤ Pu (Buldakov *et al.*, 1972; Nenot *et al.*, 1972; Efurd, 1986; NCRP, 1988), with the default assumption being that the removal half-time of neptunium is twice that of americium in cases where the assigned half-times of americium and plutonium differ. The net retention time of neptunium in a given compartment may still be shorter than that of americium in some cases because of relatively efficient elimination of neptunium from the body by urinary excretion and hence reduced recycling of neptunium.

Removal from the circulation

(B33) As in Part 1 (ICRP, 1989), the removal half-time from blood is assumed to be 0.25 days. As in the models for americium and plutonium, 30% of activity leaving blood is assigned to compartment ST0. The removal half-time from ST0 is assumed to be twice that of americium, or 1 day. Parameter values describing removal from the circulation are independent of age.

Deposition in the liver and skeleton

(B34) Assumptions concerning fractional deposition of neptunium in the liver and on trabecular and cortical bone surfaces are similar to those used in Part 1 (ICRP, 1989), except that total deposition in the skeleton has been reduced slightly to allow for greater deposition in soft tissues with intermediate turnover rates. For the adult it is assumed that 45% of neptunium leaving the circulation is deposited in the skeleton and 10% in the liver (Liver 1). By analogy with americium and plutonium, there is assumed to be elevated deposition of neptunium in the immature skeleton. Specifically, it is assumed that 70% of neptunium leaving the circulation deposits in the skeleton of the infant or 1-year-old child and 60% deposits in the skeleton of persons of ages 5–15 years. Deposition in the liver and urinary bladder contents are assumed to be reduced at younger ages due to greater competition from the skeleton. For the infant it is assumed that 3% of neptunium leaving the circulation deposits in the liver (compartment Liver 1), and for persons 5 to 15 years of age, it is assumed that 5% deposits in the liver. Deposition in the urinary bladder contents is discussed below. The initial division of neptunium between trabecular and cortical surfaces is assumed to be intermediate to those assumed for strontium and americium (ICRP, 1989). For the adult, ages 5–15 years, and age 1 year or less, it is assumed that 55%, 45% and 35%, respectively, of the deposition of neptunium in the skeleton is on trabecular surfaces and the remainder is on cortical surfaces.

Translocation of skeletal deposits

(B35) The model for translocation of skeletal deposits of neptunium is the same as that for plutonium, americium, and neptunium given in Part 1 and is summarized above in the discussion of parameter values for americium. Figure B6 gives estimated neptunium contents of skeleton as a function of time after injection of neptunium into infants, 10-year-old children, or mature adults, based on the parameter values given in Table B-3.

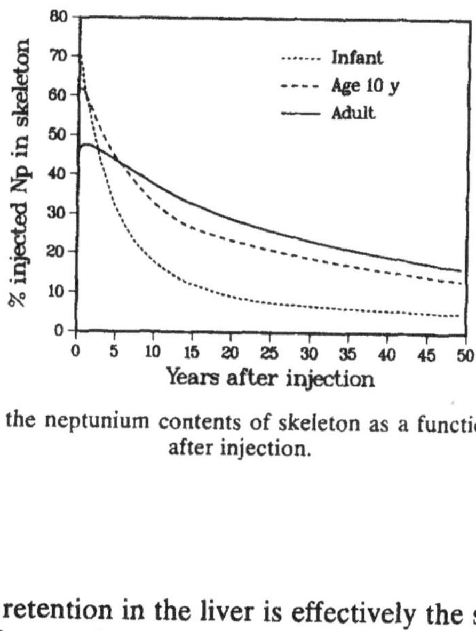

Fig. B6. Model predictions of the neptunium contents of skeleton as a function of age at injection and time after injection.

Removal from the liver

(B36) The model for retention in the liver is effectively the same as that described in Part 1 (ICRP, 1989) with regard to integrated activity, but two liver compartments are used here, as in the model for plutonium. It is assumed that neptunium depositing in the liver enters Liver 1. The removal half-time from Liver 1 is assumed to be the same as that for americium and plutonium (1 year). As in the model for plutonium, neptunium leaving Liver 1 is assumed to flow to Liver 2 except for a small amount (7% of biological removal from Liver 1, by analogy with plutonium) assigned to the gastrointestinal tract contents. Activity in Liver 2 is assumed to be removed to blood with a half-time of 1 year. The effective retention time in liver is similar to that produced by the model for neptunium in Part 1, which employed a single liver compartment with a removal half-time of 2 years. Thus, the effective removal half-time from the liver is about twice greater than that assigned to americium. Figure B7 gives estimated neptunium contents of liver as a function of time after injection of neptunium into infants, 10-year-old children, or mature adults, based on the parameter values given in Table B-3.

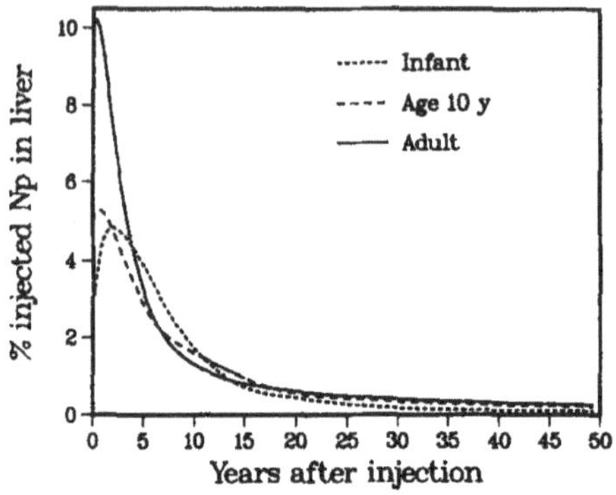

Fig. B7. Model predictions of the neptunium contents of liver as a function of age at injection and time after injection.

Uptake and removal by soft tissue compartments ST1 and ST2

(B37) As in the models for americium and plutonium, 2% of activity leaving the circulation is assigned to the soft tissue compartment ST2, from which it is removed with a half-time of 100 years. The percentage remaining after all other deposition fractions in the model have been chosen, amounting to 8.3% of neptunium leaving the circulation, is assigned to the soft tissue compartment ST1. The removal half-time from ST1 is assumed to be twice that of americium, or 100 days. Parameter values describing uptake and retention in ST1 and ST2 are independent of age.

Uptake and removal by the gonads

(B38) Assumptions concerning gonadal uptake and retention are the same as those described above for americium.

The model for urinary excretion and for retention in the kidneys

(B39) The model for urinary excretion is based on data for healthy adult human subjects (Popplewell *et al.*, 1991) and non-human primates (see ICRP, 1989) injected with neptunium. For the adult it is assumed that 32% of activity leaving the circulation goes directly to the urinary bladder contents. The rate of movement to the urinary bladder contents is assumed to be reduced at younger ages due to greater competition for neptunium by the immature skeleton. Specifically, for ages 5–15 years it is assumed that 22% of activity leaving the circulation goes to the urinary bladder contents, and for the infant and 1-year-old child it is assumed that 14% goes to the urinary bladder contents. For all ages, it is assumed that 1.5% of activity leaving the circulation deposits in the kidneys (renal tubules) and is subsequently released to the urinary bladder contents with a half-time of 14 days (twice higher than that of americium). As in the models for americium and plutonium, an additional 0.5% of activity leaving the circulation is assigned to other kidney tissues, from which it is returned to blood with a half-time of 500 days.

The model for faecal excretion

(B40) Combined data for humans and non-human primates indicate that urinary excretion is the most important route of elimination of neptunium from the body and that at most a few per cent of systemic neptunium is removed in faeces (ICRP, 1989). Reasonable agreement with faecal excretion data for non-human primates is obtained if it is assumed that fractional transfer from Liver 1 to the gastrointestinal tract via biliary secretion is the same as for plutonium, thus amounting to 7% of biological removal from Liver 1, and that biliary secretion accounts for about half of faecal excretion in the adult and direct secretion from blood into the gastrointestinal tract accounts for the other half. The latter assumption implies that 0.7% of activity leaving the circulation goes directly to the gastrointestinal contents, since the adult liver receives 10% of activity leaving the circulation. The rate of movement from blood to the gastrointestinal contents is assumed to be independent of age.

References

Atherton, D. R. and Lloyd, R. D. (1972) The distribution and retention of Cf-249 in beagle soft tissue. *Health Phys.* **22**, 675–677.

Belyaev, Yu. A. (1969) Americium-241 distribution in rats and the effect of complexing substances on its elimination. In: *Radioactive Isotopes and the Body* (Yu. I. Moskalev, Ed.), pp. 168–174. Izdatel'stvo Meditsina, Moscow. Translation AEC-tr-6944.

Breitenstein, B. D. and Palmer, H. E. (1989) Lifetime follow-up of the 1976 americium accident victim. *Radiat. Prot. Dosim.* **26**, 317–322.

Bruenger, F. W., Stevens, W. and Stover, B. J. (1969) Americium-241 in the blood: *in vivo* and *in vitro* observations. *Radiat. Res.* **37**, 349–360.

Buldakov, L. A., Nifatov, A. P., Filippova, L. G. and Khalturin, G. V. (1972) Subcutaneous administration of transuranic elements (neptunium 237, plutonium 239, and americium 241). In: *Biological Effects of Radiation from External and Internal Sources* (Yu. I. Moskalev and V. S. Kalistratova, Eds), pp. 359-367. Meditsina, Moscow.

Cohen, N. and Wrenn, M. E. (1973) Metabolic characteristics of Am-241 in the adult baboon. *Radiat. Res.* **55**, 129–143.

Cohen, N., Lo Sasso, T. and Wrenn, M. E. (1979) Metabolism of americium-241 in man: An unusual case of internal contamination of a child and his father. *Science* **206**, 64–66.

Doerfel, H. and Oliveira, C. A. N. (1989) *In vivo* measurement of Am-241 organ burdens: techniques and results of measurements after an incident in an industrial research institute. *Radiat. Prot. Dosim.* **26**, 189–193.

Durbin, P. W. (1972) Plutonium in man: a new look at the old data. In: *Radiobiology of Plutonium* (B. J. Stover and W. S. S. Jee, Eds), pp. 469–530. J. W. Press, Salt Lake City, UT.

Durbin, P. W. (1973) Metabolism and biological effects of the transplutonium elements. In: *Handbook of Experimental Pharmacology, Vol. 36: Uranium, Plutonium, and Transplutonium Elements* (H. C. Hodge, J. N. Stannard and J. B. Hursh, Eds), pp. 739–896. Springer, Berlin.

Durbin, P. W. and Schmidt, C. D. (1985) Part V: Implications for metabolic modelling. *Health Phys.* **49**, 623–661.

Durbin, P. W., Jeung, N. and Schmidt, C. T. (1985) ^{239}Pu(IV) in monkeys. Overview of metabolism. NUREG/CR-4355, Vol. 1, Lawrence Berkeley Laboratory Report 20022.

Efurd, D. W., Perrin, R. E. and McInroy, J. F. (1986) Neptunium-237 in human tissue samples. *Health Phys.* **51**, 665–666.

Fry, F. A. (1976) Long-term retention of americium-241 following accidental inhalation. *Health Phys.* **31**, 13–20.

Griffith, W. C., Mewhinney, J. A., Buggenburg, B. A., Boecker, B. B. and Cuddihy, R. G. (1983) Bioassay model for estimating body burdens of Am-241 from excretion analyses. *Health Phys.* **44**, 545–554.

Guilmette, R. A., Cohen, N. and Wrenn, M. E. (1980) Distribution and retention of Am-241 in the baboon. *Radiat. Res.* **81**, 100–119.

Heid, K. R. and Robinson, B. (1985) Part II: Estimate of the initial systemic burden. *Health Phys.* **49**, 569–575.

Hunt, G. J., Leonard, D. R. P. and Lovett, M. B. (1986) Transfer of environmental plutonium and americium across the human gut. *Sci. Total Environ.* **53**, 89.

Hunt, G. J., Leonard, D. R. P. and Lovett, M. B. (1989) Transfer of environmental plutonium and americium across the human gut. *Radiol. Prot. Bull.* **106**, 7.

Hunt, G. J., Leonard, D. R. P. and Lovett, M. B. (1990) Transfer of environmental plutonium and americium across the human gut: A second study. *Sci. Total Environ.* **90**, 273–282.

ICRP (1975) *Report of the Task Group on Reference Man*, ICRP Publication 23. Pergamon Press, Oxford.

ICRP (1986) *The Metabolism of Plutonium and Related Elements*, ICRP Publication 48. Pergamon Press, Oxford.

ICRP (1989) *Age-dependent Doses to Members of the Public from Intake of Radionuclides: Part 1*, ICRP Publication 56. Pergamon Press, Oxford.

ICRP (1991) *Annual Limits on Intake of Radionuclides by Workers based on the 1990 Recommendations*, ICRP Publication 61. Pergamon Press, Oxford.

Jones, S. R. (1985) Derivation and validation of a urinary excretion function for plutonium applicable over tens of years post uptake. *Radiat. Prot. Dosim.* **11**, 19–27.

Kathren, R. L. (1989) The United States Transuranium and Uranium Registries: overview and recent progress. *Radiat. Prot. Dosim.* **26**, 323–330.

Kathren, R. L. (1990) Distribution of plutonium in man: implications for modelling. *Health Phys.* (Abstracts of 35th Annual Meeting of the Health Physics Society), S24.

Kathren, R. L. and McInroy, J. F. (1991) Comparison of systemic plutonium deposition estimates from urinalysis and autopsy data in five whole-body donors. *Health Phys.* **60**, 481–488.

Kathren, R. L., McInroy, J. F., Reichert, M. M. and Swint, M. J. (1988) Partitioning of Pu-238, Pu-239 and Am-241 in skeleton and liver of U.S. Transuranium Registry autopsy cases. *Health Phys.* **54**, 181–188.

King, W. C. (1980) Use of AERIN code for determining internal doses of transuranic isotopes. Paper prepared for presentation at the Health Physics Society Meeting at Seattle, Washington, July 21–25, 1980; UCRL-84520 Preprint, CONF-800731—2, pp. 1–15.

Langham, W. H., Bassett, S. H., Harris, P. S. and Carter, R. E. (1950) Distribution and excretion of plutonium administered to man, Los Alamos Scientific Laboratory, LA-1151.

Leggett, R. W. (1985) A model of the retention, translocation, and excretion of systemic plutonium. *Health Phys.* **49**, 1115–1137.

Leggett, R. W. (1992) A retention-excretion model for americium in humans. *Health Phys.* **62**, 288–310.

Leggett, R. W. and Eckerman, K. F. (1987) A method for estimating the systemic burden of Pu from urinalyses. *Health Phys.* **52**, 337–346.

Lloyd, R. D., Mays, C. W., Taylor, G. N. and Atherton, D. R. (1970) Americium-241 studies in beagles. *Health Phys.* **18**, 149–156.

Lloyd, R. D., Jee, W. S. S., Atherton, D. R., Taylor, G. N. and Mays, C. W. (1972) Americium-241 in beagles: Biological effects and skeletal distribution. In: *Radiobiology of Plutonium* (B. J. Stover and W. S. S. Jee, Eds), pp. 141–148. J. W. Press, Salt Lake City, UT.

Lloyd, R. D., Atherton, D. R., Mays, C. W., McFarland, S. S. and Williams, J. L. (1974) The early excretion, retention and distribution of injected curium citrate in beagles. *Health Phys.* **27**, 61–67.

Lloyd, R. D., Atherton, D. R., McFarland, S. S., Mays, C. W., Stevens, W., Williams, J. L. and Taylor, G. N. (1976) Studies of Pu(IV) citrate in beagles. *Health Phys.* **30**, 47–52.

Lloyd, R. D., McFarland, S. S., Atherton, D. R. and Mays, C. W. (1978a) Plutonium retention, excretion, and distribution in juvenile beagles soon after injection. *Radiat. Res.* **75**, 633–641.

Lloyd, R. D., McFarland, S. S., Atherton, D. R., Bruenger, F. W., Taylor, G. N. and Mays, C. W. (1978b) Early retention of Pu-237 + Pu-239 in mature beagles. *Health Phys.* **35**, 211–215.

Lloyd, R. D., Mays, C. W., Jones, C. W., Atherton, D. R., Bruenger, F. W., Shabestari, L. R. and Wrenn, M. E. (1984) Retention and dosimetry of injected Am-241 in beagles. *Radiat. Res.* **110**, 564–575.

Maletskos, C. J., Kerne, A. T., Telles, N. C. and Evans, R. D. (1969) Retention and absorption of 224-Ra and 234-Th and some dosimetric consequences in human beings. In: *Delayed Effects of Bone Seeking Radionuclides* (C. W. Mays, W. S. S. Jee and R. D. Lloyd, Eds). University of Utah Press, Salt Lake City, UT.

McClellan, R. O., Casey, H. W., Bustad, L. K. (1962) Transfer of some transuranic elements to milk. *Health Phys.* **8**, 689–694.

McInroy, J. F. and Kathren, R. L. (1990) Plutonium content in marrow and mineralized bone in an occupationally exposed person. *Radiat. Prot. Dosim.* **32**, 245–252.

McInroy, J. F., Kathren, R. L. and Swint, M. J. (1989) Distribution of plutonium and americium in whole bodies donated to the United States Transuranium Registry. *Radiat. Prot. Dosim.* **26**, 151–158.

McKay, L. R., Brooks, A. L. and McClellan, R. O. (1972) The retention, distribution, dose, and cytogenetic effects of Am-241 citrate in the chinese hamster. *Health Phys.* **22**, 633–640.

Miller, S. C., Bruenger, F. W. and Williams, F. W. (1989) Influence of age at exposure on concentrations of Pu-239 in beagle gonads. *Health Phys.* **56**, 485–491.

Moskalev, Y. I. (1977) *Problems of the Radiobiology of Americium-241.* DOE-tr-190, prepared by the U.S. Joint Publications Service, Arlington, VA, 1979, translated from the Russian book *Problemy Radiobiologii Ameritsiya-241.* Atomizdat.

Moss, W. D., Tietjen, G. L., Gautier, M. A. and McLeod, M. J. (1983) A reivew of the human plutonium injection studies. Proceedings of the 29th Annual Conference on Bioassay, Analytical and Environmental Chemistry, Seattle, Washington, Oct. 12–13, 1983 (CONF-8310157, order no. DE83017911, available from National Technical Information Service, U.S. Department of Commerce, 5285 Port Royal Road, Springfield, VA 22161).

Mussalo-Rauhamaa, H., Jaakola, T., Miettinen, J. K. and Laiho, K. (1984) Plutonium in Finnish Lapps—An estimate of the gastrointestinal absorption of plutonium by man based on a comparison of the plutonium content of Lapps and southern Finns. *Health Phys.* **46**, 549.

NCRP (1989) National Council on Radiation Protection and Measurements. *Neptunium: Radiation Protection Considerations*. NCRP Report No. 90.

NEA/OECD (1988) Committee on Radiation Protection and Public Health. *Report of an Expert Group on Gut Transfer Factors*. NEA/OECD Report, Paris.

Nenot, J. C., Masse, R., Morin, M., Lafuma, J. (1972) An experimental comparative study of the behaviour of Np-237, Pu-238, Pu-239, Am-241 and Cm-242 in bone. *Health Phys.* **22**, 657–665.

Popplewell, D. S. and Ham, G. J. (1989) Distribution of plutonium and americium in tissues from a human autopsy case. *J. Radiol. Prot.* **9**, 159–164.

Popplewell, D. S., Harrison, J. D. and Ham, G. J. (1989) The measurement of the gastrointestinal absorption of neptunium-239 and curium-242 in humans. Paper presented at the 4th Int. Symp., Radiation Protection—Theory and Practice, Malvern, June 1989.

Popplewell, D. S., Harrison, J. D. and Ham, G. J. (1991) The gastrointestinal absorption of neptunium and curium in humans. *Health Phys.* **60**, 797–805.

Robinson, B., Heid, K. R., Aldridge, T. L. and Glenn, R. D. (1983) 1976 Hanford americium exposure incident: Organ burden and radiation dose estimates. *Health Phys.* **45**, 911–921.

Rosen, J. C., Cohen, N. and Wrenn, M. E. (1972) Short term metabolism of Am-241 in the adult baboon. *Health Phys.* **22**, 621–626.

Rosen, J. C., Gur, D., Pan, S. F., Wald, N. and Brodsky, A. (1980) Long-term removal of Am-241 using Ca-DTPA. *Health Phys.* **39**, 601–609.

Rundo, J. (1979) The late excretion of plutonium following acquisition of known amounts. In: *Actinides in Man and Animals. Proceedings of the Snowbird Actinide Workshop* (M. E. Wrenn, Ed.), pp. 253–260.

Rundo, J., Starzyk, P. M., Sedlet, J., Larsen, R. P., Oldham, R. D. and Robinson, J. J. (1976) The excretion rate and retention of plutonium 10,000 days after acquisition. In: *Diagnosis and Treatment of Incorporated Radionuclides*, pp. 15–22. IAEA, Vienna.

Scott, K. G., Copp, D. H., Axelrod, D. J. and Hamilton, J. G. (1948) The metabolism of americium in the rat. *J. Biol. Chem.* **175**, 691–705.

Stevens, W. and Bruenger, F. W. (1972) Interaction of Cf-249 and Cf-252 with constituents of dog and human blood. *Health Phys.* **22**, 679–683.

Sutton, W. W., Patzer, R. G., Mullen, A. A., Hahn, P. B. and Potter, G. D. (1978) Metabolism of americium-241 in dairy animals. In *Selected Environmental Plutonium Research Reports of the Nevada Applied Ecology Group* (M. G. White and P. B. Dunaway, Eds), Report NVO-192, pp. 19–43.

Takada, K., Fukuda, H., Hattori, T. and Akaishi, J. (1984) Equation for the urinary excretion of americium following a single uptake. In: *Radiation—Risk—Protection, Compacts*, Vol. I (A. Kaul, R. Neider, J. Pensko, F.-E. Stieve and H. Brunner, Eds), pp. 477–480. 6th International Congress organized by the Fachverband für Strahlenschutz e.V. Berlin (West) May 7–12, 1984; IRPA.

Talbot, R. J., Newton, D. and Warner, A. J. (1992) Personal communication to Task Group. Harwell Biomedical Research, Harwell Laboratory, Oxfordshire, UK.

Toohey, R. E. and Essling, M. A. (1980) Measurements of Am-241 *in vivo* at long times after inhalation. *Health Phys.* **38**, 139–145.

Turner, G. A. and Taylor, D. M. (1968) The transport of plutonium, americium and curium in the blood of rats. *Phys. Med. Biol.* **13**, 535–546.

Voelz, G. L., Hempelmann, L. H., Lawrence, J. N. P. and Moss, W. D. A. (1979) 32-year medical follow-up of Manhattan Project plutonium workers. *Health Phys.* **37**, 445–485.

Voelz, G. L., Grier, R. S. and Hempelmann, L. H. A. (1985) 37-year medical follow-up of Manhattan Project Pu workers. *Health Phys.* **48**, 249–259.

Whalen, R. P. and Davies, S. (1972) Americium contamination incident in a New York State Health Department laboratory. Radiation and Data Reports, May 1972, 249–253.

Wrenn, M. E., Rosen, J. C. and Cohen, N. (1972) *In-vivo* measurement of americium-24 in man. In: *Assessment of Radioactive Contamination in Man, Proceedings of a Symposium Organized by IAEA and WHO, Stockholm, 22–26 November 1971*, pp. 595–621. IAEA-SM-150/38. IAEA, Vienna.

Wright, S. R. L. (1972) X-ray measurement of actinide metabolism using Am-243 in mice. *Int. J. Appl. Radiat. Isot.* **23**, 423–429.

Zalikin, G. A., Moskalev, Yu. I. and Petrovich, I. K. (1968) Distribution and biological effects of americium-241. *Radiobiology* **8**, 107–118 (AEC-tr-6950).

APPENDIX C: AGE-SPECIFIC BIOKINETIC DATA AND DOSE COEFFICIENTS FOR RADIOISOTOPES OF HYDROGEN, CARBON, ZIRCONIUM, NIOBIUM, RUTHENIUM, IODINE, CAESIUM, CERIUM, PLUTONIUM, AMERICIUM AND NEPTUNIUM

(C1) This appendix gives revised ingestion dose coefficients for the radioisotopes of the elements in Part 1 (ICRP, 1989), excluding strontium, using the new tissue weighting factors recommended by ICRP (1991). Dose coefficients are given for the following radionuclides:

^{3}H; ^{14}C; ^{95}Zr; ^{95}Nb; ^{103}Ru; ^{106}Ru; ^{129}I; ^{131}I; ^{132}I; ^{134}Cs; ^{137}Cs; ^{144}Ce; ^{238}Pu; ^{239}Pu; ^{241}Am; ^{237}Np; ^{239}Np.

(C2) In addition, ingestion dose coefficients for the following radionuclides that were not included in Part 1 are given here:

^{125}I; ^{136}Cs; ^{141}Ce; ^{240}Pu.

(C3) For the purpose of calculating doses to the urinary bladder, urinary to faecal excretion ratios are now included.

(C4) It has not been necessary to apply the general approach for assessing appropriate f_1 values described in paragraphs (10) and (11) in the Introduction as sufficient specific information is available for those elements in Part 1 that are readily absorbed from the gastrointestinal tract.

(C5) No reference is now made to biokinetic data given in previous ICRP publications; all necessary data are given in this report.

(C6) Tables of biokinetic data are included here together with dose coefficients. In Part 1 the dose coefficients for ^{144}Ce were computed as if cerium was volume distributed in the skeleton. Cerium isotopes should be assumed to be distributed on bone surfaces as considered in this report.

Table C-1.1. Biokinetic data for tritiated water

		Distribution (%) total body		Biological half-time (days)	
Age	f_1	Comp. A	Comp. B	Comp. A	Comp. B
3 months	1	97	3	3.0	8
1 year	1	97	3	3.5	15
5 years	1	97	3	4.6	19
10 years	1	97	3	5.7	26
15 years	1	97	3	7.9	32
Adult	1	97	3	10.0	40

It is assumed that tritiated water is distributed instantaneously over the total body without additional nuclear transformations within the stomach.

The concentration in urine is assumed to be the same as in total body water. Urinary excretion will not result in a significant additional dose to the bladder wall which is assumed to receive the same dose as other tissues.

Table C-1.2

Ingestion Dose Coefficients (Equivalent Dose (Sv/Bq) to Age 70 Years) for Tritiated Water (H-3 T1/2 - 12.35 y)

Age at intake	3 Months	1 Year	5 Years	10 Years	15 Years	Adult
Adrenals	6.3E-11	4.8E-11	3.0E-11	2.3E-11	1.8E-11	1.8E-11
Bladder Wall	6.3E-11	4.8E-11	3.0E-11	2.3E-11	1.8E-11	1.8E-11
Bone Surfaces	6.3E-11	4.8E-11	3.0E-11	2.3E-11	1.8E-11	1.8E-11
Brain	6.3E-11	4.8E-11	3.0E-11	2.3E-11	1.8E-11	1.8E-11
Breast	6.3E-11	4.8E-11	3.0E-11	2.3E-11	1.8E-11	1.8E-11
GI-Tract						
St Wall	6.3E-11	4.8E-11	3.0E-11	2.3E-11	1.8E-11	1.8E-11
SI Wall	6.3E-11	4.8E-11	3.0E-11	2.3E-11	1.8E-11	1.8E-11
ULI Wall	6.3E-11	4.8E-11	3.0E-11	2.3E-11	1.8E-11	1.8E-11
LLI Wall	6.3E-11	4.8E-11	3.0E-11	2.3E-11	1.8E-11	1.8E-11
Kidneys	6.3E-11	4.8E-11	3.0E-11	2.3E-11	1.8E-11	1.8E-11
Liver	6.3E-11	4.8E-11	3.0E-11	2.3E-11	1.8E-11	1.8E-11
Lungs	6.3E-11	4.8E-11	3.0E-11	2.3E-11	1.8E-11	1.8E-11
Muscle	6.3E-11	4.8E-11	3.0E-11	2.3E-11	1.8E-11	1.8E-11
Ovaries	6.3E-11	4.8E-11	3.0E-11	2.3E-11	1.8E-11	1.8E-11
Pancreas	6.3E-11	4.8E-11	3.0E-11	2.3E-11	1.8E-11	1.8E-11
Red Marrow	6.3E-11	4.8E-11	3.0E-11	2.3E-11	1.8E-11	1.8E-11
Skin	6.3E-11	4.8E-11	3.0E-11	2.3E-11	1.8E-11	1.8E-11
Spleen	6.3E-11	4.8E-11	3.0E-11	2.3E-11	1.8E-11	1.8E-11
Testes	6.3E-11	4.8E-11	3.0E-11	2.3E-11	1.8E-11	1.8E-11
Thymus	6.3E-11	4.8E-11	3.0E-11	2.3E-11	1.8E-11	1.8E-11
Thyroid	6.3E-11	4.8E-11	3.0E-11	2.3E-11	1.8E-11	1.8E-11
Uterus	6.3E-11	4.8E-11	3.0E-11	2.3E-11	1.8E-11	1.8E-11
Remainder	6.3E-11	4.8E-11	3.0E-11	2.3E-11	1.8E-11	1.8E-11
Effective Dose	6.3E-11	4.8E-11	3.0E-11	2.3E-11	1.8E-11	1.8E-11

GI-Tract Gastrointestinal Tract
St Stomach
SI Small Intestine
ULI Upper Large Intestine
LLI Lower Large Intestine

Table C-2.1. Biokinetic data for organically bound tritium

		Distribution (%) total body		Biological half-time (days)	
Age	f_1	Comp. A	Comp. B	Comp. A	Comp. B
3 months	1	50	50	3.0	8
1 year	1	50	50	3.5	15
5 years	1	50	50	4.6	19
10 years	1	50	50	5.7	26
15 years	1	50	50	7.9	32
Adult	1	50	50	10.0	40

In contrast to tritiated water additional nuclear transformations within the stomach due to the biokinetic gastrointestinal tract model have been assumed.

The concentration in urine is assumed to be the same as in total body water. Urinary excretion will not result in a significant additional dose to the bladder wall which is assumed to receive the same dose as other tissues.

Table C-2.2

Ingestion Dose Coefficients (Equivalent Dose (Sv/Bq) to Age 70 Years) for Organically Bound Tritium (H-3 T1/2 = 12.35 y)

Age at intake	3 Months	1 Year	5 Years	10 Years	15 Years	Adult
Adrenals	1.1E-10	1.1E-10	6.9E-11	5.5E-11	4.0E-11	4.1E-11
Bladder Wall	1.1E-10	1.1E-10	6.9E-11	5.5E-11	4.0E-11	4.1E-11
Bone Surfaces	1.1E-10	1.1E-10	6.9E-11	5.5E-11	4.0E-11	4.1E-11
Brain	1.1E-10	1.1E-10	6.9E-11	5.5E-11	4.0E-11	4.1E-11
Breast	1.1E-10	1.1E-10	6.9E-11	5.5E-11	4.0E-11	4.1E-11
GI-Tract						
St Wall	1.9E-10	1.6E-10	9.1E-11	6.8E-11	4.9E-11	4.8E-11
SI Wall	1.1E-10	1.1E-10	7.0E-11	5.5E-11	4.1E-11	4.1E-11
ULI Wall	1.2E-10	1.2E-10	7.3E-11	5.7E-11	4.2E-11	4.2E-11
LLI Wall	1.4E-10	1.3E-10	8.0E-11	6.1E-11	4.4E-11	4.4E-11
Kidneys	1.1E-10	1.1E-10	6.9E-11	5.5E-11	4.0E-11	4.1E-11
Liver	1.1E-10	1.1E-10	6.9E-11	5.5E-11	4.0E-11	4.1E-11
Lungs	1.1E-10	1.1E-10	6.9E-11	5.5E-11	4.0E-11	4.1E-11
Muscle	1.1E-10	1.1E-10	6.9E-11	5.5E-11	4.0E-11	4.1E-11
Ovaries	1.1E-10	1.1E-10	6.9E-11	5.5E-11	4.0E-11	4.1E-11
Pancreas	1.1E-10	1.1E-10	6.9E-11	5.5E-11	4.0E-11	4.1E-11
Red Marrow	1.1E-10	1.1E-10	6.9E-11	5.5E-11	4.0E-11	4.1E-11
Skin	1.1E-10	1.1E-10	6.9E-11	5.5E-11	4.0E-11	4.1E-11
Spleen	1.1E-10	1.1E-10	6.9E-11	5.5E-11	4.0E-11	4.1E-11
Testes	1.1E-10	1.1E-10	6.9E-11	5.5E-11	4.0E-11	4.1E-11
Thymus	1.1E-10	1.1E-10	6.9E-11	5.5E-11	4.0E-11	4.1E-11
Thyroid	1.1E-10	1.1E-10	6.9E-11	5.5E-11	4.0E-11	4.1E-11
Uterus	1.1E-10	1.1E-10	6.9E-11	5.5E-11	4.0E-11	4.1E-11
Remainder	1.1E-10	1.1E-10	6.9E-11	5.5E-11	4.0E-11	4.1E-11
Effective Dose	1.2E-10	1.2E-10	7.3E-11	5.7E-11	4.2E-11	4.2E-11

GI-Tract Gastrointestinal Tract
St Stomach
SI Small Intestine
ULI Upper Large Intestine
LLI Lower Large Intestine

Table C-3.1. Biokinetic data for carbon

Age	f_1	Distribution (%) total body	Biological half-time (days)
3 months	1	100	8
1 year	1	100	15
5 years	1	100	19
10 years	1	100	26
15 years	1	100	32
Adult	1	100	40

Excretion will not result in a significant additional dose to the bladder wall and lower and upper large intestine wall which are assumed to receive the same dose as other tissues.

Table C-3.2

Ingestion Dose Coefficients (Equivalent Dose (Sv/Bq) to Age 70 Years) for C-14 (T1/2 = 5730 y)

Age at intake	3 Months	1 Year	5 Years	10 Years	15 Years	Adult
Adrenals	1.3E-09	1.6E-09	9.7E-10	7.8E-10	5.7E-10	5.7E-10
Bladder Wall	1.3E-09	1.6E-09	9.7E-10	7.8E-10	5.7E-10	5.7E-10
Bone Surfaces	1.3E-09	1.6E-09	9.7E-10	7.8E-10	5.7E-10	5.7E-10
Brain	1.3E-09	1.6E-09	9.7E-10	7.8E-10	5.7E-10	5.7E-10
Breast	1.3E-09	1.6E-09	9.7E-10	7.8E-10	5.7E-10	5.7E-10
GI-Tract						
St Wall	2.1E-09	2.0E-09	1.2E-09	8.9E-10	6.4E-10	6.3E-10
SI Wall	1.3E-09	1.6E-09	9.7E-10	7.9E-10	5.7E-10	5.7E-10
ULI Wall	1.4E-09	1.6E-09	1.0E-09	8.0E-10	5.8E-10	5.8E-10
LLI Wall	1.6E-09	1.7E-09	1.1E-09	8.4E-10	6.0E-10	6.0E-10
Kidneys	1.3E-09	1.6E-09	9.7E-10	7.8E-10	5.7E-10	5.7E-10
Liver	1.3E-09	1.6E-09	9.7E-10	7.8E-10	5.7E-10	5.7E-10
Lungs	1.3E-09	1.6E-09	9.7E-10	7.8E-10	5.7E-10	5.7E-10
Muscle	1.3E-09	1.6E-09	9.7E-10	7.8E-10	5.7E-10	5.7E-10
Ovaries	1.3E-09	1.6E-09	9.7E-10	7.8E-10	5.7E-10	5.7E-10
Pancreas	1.3E-09	1.6E-09	9.7E-10	7.8E-10	5.7E-10	5.7E-10
Red Marrow	1.3E-09	1.6E-09	9.7E-10	7.8E-10	5.7E-10	5.7E-10
Skin	1.3E-09	1.6E-09	9.7E-10	7.8E-10	5.7E-10	5.7E-10
Spleen	1.3E-09	1.6E-09	9.7E-10	7.8E-10	5.7E-10	5.7E-10
Testes	1.3E-09	1.6E-09	9.7E-10	7.8E-10	5.7E-10	5.7E-10
Thymus	1.3E-09	1.6E-09	9.7E-10	7.8E-10	5.7E-10	5.7E-10
Thyroid	1.3E-09	1.6E-09	9.7E-10	7.8E-10	5.7E-10	5.7E-10
Uterus	1.3E-09	1.6E-09	9.7E-10	7.8E-10	5.7E-10	5.7E-10
Remainder	1.3E-09	1.6E-09	9.7E-10	7.8E-10	5.7E-10	5.7E-10
Effective Dose	1.4E-09	1.6E-09	1.0E-09	8.0E-10	5.8E-10	5.8E-10

GI-Tract Gastrointestinal Tract
St Stomach
SI Small Intestine
ULI Upper Large Intestine
LLI Lower Large Intestine

Excretion of systemic C-14 was not routed through the urinary bladder and the gastrointestinal tract, but this does not significantly effect the estimated dose for these tissues.

Table C-4.1. Biokinetic data for zirconium

Age	f_1	Distribution (%)		Biological half-time (days)	
		Skeleton	Other tissues	Skeleton*	Other tissues
3 months	$2 \cdot 10^{-2}$	50	50	100	7
1 year	$1 \cdot 10^{-2}$	50	50	300	7
5 years	$1 \cdot 10^{-2}$	50	50	500	7
10 years	$1 \cdot 10^{-2}$	50	50	1000	7
15 years	$1 \cdot 10^{-2}$	50	50	1500	7
Adult	$1 \cdot 10^{-2}$	50	50	10 000	7

There appears to be no specific information on faecal excretion of zirconium. Due to its chemical similarities to niobium, zirconium that has entered the transfer compartment is assumed to have a urinary to faecal excretion ratio of 5:1.

*Half-time modified from 8000 days given in Part 1 according to the general approach described in paragraph (7) of the Introduction.

Table C-4.2

Ingestion Dose Coefficients (Equivalent Dose (Sv/Bq) to Age 70 Years) for Zr-95 (T1/2 = 63.98 d)

Age at intake	3 Months	1 Year	5 Years	10 Years	15 Years	Adult
Adrenals	9.1E-10	5.8E-10	3.4E-10	2.3E-10	1.6E-10	1.3E-10
Bladder Wall	2.0E-09	1.5E-09	8.2E-10	6.1E-10	3.4E-10	2.9E-10
Bone Surfaces	1.5E-08	8.2E-09	6.2E-09	3.6E-09	2.2E-09	2.3E-09
Brain	3.6E-10	1.8E-10	1.1E-10	8.1E-11	6.0E-11	5.3E-11
Breast	3.6E-10	2.1E-10	1.1E-10	7.6E-11	4.3E-11	3.7E-11
GI-Tract						
St Wall	3.5E-09	2.1E-09	1.2E-09	7.3E-10	4.8E-10	3.9E-10
SI Wall	8.3E-09	5.9E-09	3.3E-09	2.1E-09	1.3E-09	1.1E-09
ULI Wall	3.0E-08	2.0E-08	1.1E-08	6.6E-09	3.9E-09	3.1E-09
LLI Wall	7.8E-08	5.2E-08	2.7E-08	1.6E-08	9.6E-09	7.8E-09
Kidneys	1.0E-09	6.7E-10	4.3E-10	2.9E-10	2.0E-10	1.7E-10
Liver	9.8E-10	6.4E-10	3.7E-10	2.3E-10	1.3E-10	1.1E-10
Lungs	5.0E-10	2.9E-10	1.6E-10	1.1E-10	7.3E-11	6.0E-11
Muscle	9.3E-10	6.2E-10	3.7E-10	2.5E-10	1.7E-10	1.4E-10
Ovaries	5.4E-09	4.1E-09	2.4E-09	1.7E-09	1.2E-09	8.8E-10
Pancreas	1.1E-09	7.0E-10	4.2E-10	2.8E-10	1.8E-10	1.5E-10
Red Marrow	4.2E-09	2.0E-09	1.2E-09	8.3E-10	5.8E-10	4.8E-10
Skin	5.3E-10	3.1E-10	1.8E-10	1.2E-10	7.5E-11	6.3E-11
Spleen	9.2E-10	5.8E-10	3.4E-10	2.2E-10	1.4E-10	1.1E-10
Testes	9.9E-10	6.7E-10	3.7E-10	2.3E-10	1.3E-10	1.0E-10
Thymus	3.3E-10	1.7E-10	1.1E-10	7.0E-11	5.0E-11	4.3E-11
Thyroid	2.8E-10	1.5E-10	8.9E-11	6.3E-11	4.7E-11	4.3E-11
Uterus	2.8E-09	2.1E-09	1.2E-09	8.1E-10	5.1E-10	4.0E-10
Remainder	9.7E-10	6.3E-10	3.9E-10	2.7E-10	1.9E-10	1.5E-10
Effective Dose	8.5E-09	5.7E-09	3.1E-09	1.9E-09	1.2E-09	9.6E-10

GI-Tract	Gastrointestinal Tract
St	Stomach
SI	Small Intestine
ULI	Upper Large Intestine
LLI	Lower Large Intestine

Table C-5.1. Biokinetic data for niobium

		Distribution (%)				Biological half-time (days)	
Age	f_1	Skeleton	Liver	Kidneys	Other tissues	Comp. A 50%	Comp. B 50%
3 months	$2 \cdot 10^{-2}$	60	15	2	23	6	200
1 year	$1 \cdot 10^{-2}$	50	18	2	30	6	200
5 years	$1 \cdot 10^{-2}$	50	18	2	30	6	200
10 years	$1 \cdot 10^{-2}$	40	20	3	37	6	200
15 years	$1 \cdot 10^{-2}$	40	20	3	37	6	200
Adult	$1 \cdot 10^{-2}$	40	20	3	37	6	200

A urinary to faecal excretion ratio of 5:1 is assumed for niobium that has entered the transfer compartment.

Table C-5.2

Ingestion Dose Coefficients (Equivalent Dose (Sv/Bq) to Age 70 Years) for Nb-95 (T1/2 = 35.15 d)

Age at intake	3 Months	1 Year	5 Years	10 Years	15 Years	Adult
Adrenals	6.9E-10	4.3E-10	2.4E-10	1.5E-10	1.0E-10	7.7E-11
Bladder Wall	1.9E-09	1.4E-09	7.7E-10	5.6E-10	3.1E-10	2.6E-10
Bone Surfaces	2.8E-09	1.2E-09	7.3E-10	3.7E-10	2.3E-10	2.1E-10
Brain	1.5E-10	5.9E-11	3.3E-11	2.1E-11	1.3E-11	1.1E-11
Breast	2.6E-10	1.5E-10	7.7E-11	4.8E-11	2.5E-11	2.0E-11
GI-Tract						
St Wall	2.4E-09	1.5E-09	8.5E-10	5.5E-10	3.6E-10	2.9E-10
SI Wall	5.8E-09	4.2E-09	2.5E-09	1.6E-09	1.0E-09	8.3E-10
ULI Wall	1.5E-08	1.1E-08	5.8E-09	3.7E-09	2.3E-09	1.8E-09
LLI Wall	3.5E-08	2.4E-08	1.3E-08	8.1E-09	4.9E-09	4.0E-09
Kidneys	1.1E-09	6.5E-10	4.0E-10	2.9E-10	1.9E-10	1.6E-10
Liver	1.3E-09	7.9E-10	4.5E-10	2.9E-10	1.8E-10	1.4E-10
Lungs	3.6E-10	2.1E-10	1.0E-10	6.5E-11	4.1E-11	3.0E-11
Muscle	7.7E-10	5.2E-10	3.0E-10	2.0E-10	1.3E-10	1.1E-10
Ovaries	5.2E-09	4.0E-09	2.3E-09	1.6E-09	1.1E-09	8.2E-10
Pancreas	1.0E-09	6.5E-10	3.8E-10	2.5E-10	1.5E-10	1.2E-10
Red Marrow	1.2E-09	5.7E-10	4.1E-10	3.0E-10	2.2E-10	1.9E-10
Skin	3.8E-10	2.3E-10	1.3E-10	8.0E-11	5.2E-11	4.3E-11
Spleen	8.0E-10	5.1E-10	2.9E-10	1.9E-10	1.1E-10	9.0E-11
Testes	8.8E-10	6.1E-10	3.3E-10	2.0E-10	1.1E-10	8.9E-11
Thymus	2.1E-10	1.1E-10	6.4E-11	3.7E-11	2.4E-11	1.9E-11
Thyroid	1.7E-10	8.5E-11	4.2E-11	2.7E-11	1.6E-11	1.3E-11
Uterus	2.7E-09	2.0E-09	1.2E-09	7.6E-10	4.7E-10	3.7E-10
Remainder	7.3E-10	4.9E-10	3.0E-10	2.1E-10	1.4E-10	1.2E-10
Effective Dose	4.6E-09	3.2E-09	1.8E-09	1.2E-09	7.4E-10	5.9E-10

GI-Tract Gastrointestinal Tract
St Stomach
SI Small Intestine
ULI Upper Large Intestine
LLI Lower Large Intestine

Table C-6.1. Biokinetic data for ruthenium

		Distribution (%)				Biological half-time (days)			
						Total body			
Age	f_1	Comp. A	Comp. B	Comp. C	Prompt excretion	Comp. A	Comp. B	Comp. C	Transfer Comp.
3 months	0.1	35	30	20	15	8	35	1000	0.3
1 year	0.05	35	30	20	15	8	35	1000	0.3
5 years	0.05	35	30	20	15	8	35	1000	0.3
10 years	0.05	35	30	20	15	8	35	1000	0.3
15 years	0.05	35	30	20	15	8	35	1000	0.3
Adults	0.05	35	30	20	15	8	35	1000	0.3

A urinary to faecal excretion ratio of 4:1 is assumed for ruthenium that has entered the transfer compartment.

Table C-6.2

Ingestion Dose Coefficients (Equivalent Dose (Sv/Bq) to Age 70 Years) for Ru-103 (T1/2 = 39.28 d)

Age at intake	3 Months	1 Year	5 Years	10 Years	15 Years	Adult
Adrenals	1.3E-09	5.9E-10	3.2E-10	2.0E-10	1.3E-10	1.0E-10
Bladder Wall	2.1E-09	1.2E-09	7.0E-10	4.9E-10	2.8E-10	2.4E-10
Bone Surfaces	1.3E-09	5.9E-10	3.1E-10	2.0E-10	1.3E-10	1.1E-10
Brain	9.7E-10	3.5E-10	1.9E-10	1.1E-10	6.9E-11	5.8E-11
Breast	9.3E-10	3.6E-10	1.9E-10	1.1E-10	7.2E-11	5.9E-11
GI-Tract						
St Wall	3.3E-09	1.8E-09	9.4E-10	5.7E-10	3.9E-10	3.1E-10
SI Wall	6.9E-09	4.6E-09	2.5E-09	1.6E-09	9.9E-10	8.0E-10
ULI Wall	2.6E-08	1.8E-08	9.0E-09	5.6E-09	3.2E-09	2.6E-09
LLI Wall	6.6E-08	4.5E-08	2.3E-08	1.4E-08	8.2E-09	6.6E-09
Kidneys	1.4E-09	6.6E-10	3.9E-10	2.5E-10	1.6E-10	1.4E-10
Liver	1.5E-09	7.0E-10	3.8E-10	2.3E-10	1.4E-10	1.1E-10
Lungs	1.1E-09	4.4E-10	2.3E-10	1.4E-10	9.0E-11	7.2E-11
Muscle	1.3E-09	6.3E-10	3.5E-10	2.2E-10	1.5E-10	1.2E-10
Ovaries	4.2E-09	2.9E-09	1.7E-09	1.1E-09	7.7E-10	5.7E-10
Pancreas	1.6E-09	7.4E-10	4.2E-10	2.7E-10	1.7E-10	1.3E-10
Red Marrow	1.2E-09	5.6E-10	3.7E-10	2.7E-10	1.9E-10	1.6E-10
Skin	9.8E-10	4.0E-10	2.1E-10	1.3E-10	8.3E-11	7.0E-11
Spleen	1.4E-09	6.4E-10	3.6E-10	2.3E-10	1.4E-10	1.2E-10
Testes	1.4E-09	7.1E-10	3.6E-10	2.3E-10	1.3E-10	1.1E-10
Thymus	1.1E-09	4.0E-10	2.2E-10	1.3E-10	8.6E-11	7.1E-11
Thyroid	1.1E-09	4.1E-10	2.2E-10	1.3E-10	8.2E-11	6.7E-11
Uterus	2.6E-09	1.6E-09	9.4E-10	6.0E-10	3.8E-10	3.0E-10
Remainder	1.4E-09	6.6E-10	3.7E-10	2.4E-10	1.6E-10	1.3E-10
Effective Dose	7.1E-09	4.6E-09	2.4E-09	1.5E-09	9.2E-10	7.3E-10

GI-Tract Gastrointestinal Tract
St Stomach
SI Small Intestine
ULI Upper Large Intestine
LLI Lower Large Intestine

Table C-6.3

Ingestion Dose Coefficients (Equivalent Dose (Sv/Bq) to Age 70 Years) for Ru-106 (T1/2 = 368.2 d)

Age at intake	3 Months	1 Year	5 Years	10 Years	15 Years	Adult
Adrenals	2.3E-08	8.8E-09	4.8E-09	2.9E-09	1.8E-09	1.5E-09
Bladder Wall	2.4E-08	9.3E-09	5.3E-09	3.3E-09	2.1E-09	1.7E-09
Bone Surfaces	2.3E-08	8.7E-09	4.7E-09	2.8E-09	1.7E-09	1.5E-09
Brain	2.3E-08	8.6E-09	4.7E-09	2.8E-09	1.7E-09	1.4E-09
Breast	2.3E-08	8.6E-09	4.6E-09	2.8E-09	1.7E-09	1.4E-09
GI-Tract						
St Wall	4.4E-08	2.0E-08	1.0E-08	6.0E-09	3.9E-09	3.2E-09
SI Wall	6.8E-08	3.9E-08	2.0E-08	1.2E-08	6.8E-09	5.6E-09
ULI Wall	2.9E-07	1.9E-07	9.6E-08	5.7E-08	3.2E-08	2.6E-08
LLI Wall	7.9E-07	5.3E-07	2.6E-07	1.6E-07	8.9E-08	7.2E-08
Kidneys	2.3E-08	8.8E-09	4.8E-09	2.9E-09	1.8E-09	1.5E-09
Liver	2.3E-08	8.8E-09	4.8E-09	2.9E-09	1.8E-09	1.5E-09
Lungs	2.3E-08	8.7E-09	4.7E-09	2.8E-09	1.7E-09	1.4E-09
Muscle	2.3E-08	8.7E-09	4.7E-09	2.8E-09	1.7E-09	1.5E-09
Ovaries	2.4E-08	9.8E-09	5.4E-09	3.3E-09	2.0E-09	1.7E-09
Pancreas	2.3E-08	8.9E-09	4.8E-09	2.9E-09	1.8E-09	1.5E-09
Red Marrow	2.3E-08	8.7E-09	4.8E-09	2.9E-09	1.8E-09	1.5E-09
Skin	2.3E-08	8.6E-09	4.6E-09	2.8E-09	1.7E-09	1.4E-09
Spleen	2.3E-08	8.8E-09	4.8E-09	2.9E-09	1.8E-09	1.5E-09
Testes	2.3E-08	8.8E-09	4.7E-09	2.8E-09	1.7E-09	1.5E-09
Thymus	2.3E-08	8.7E-09	4.7E-09	2.8E-09	1.7E-09	1.4E-09
Thyroid	2.3E-08	8.7E-09	4.7E-09	2.8E-09	1.7E-09	1.4E-09
Uterus	2.4E-08	9.2E-09	5.0E-09	3.0E-09	1.9E-09	1.6E-09
Remainder	2.4E-08	9.4E-09	5.1E-09	3.0E-09	1.8E-09	1.5E-09
Effective Dose	8.4E-08	5.0E-08	2.5E-08	1.5E-08	8.6E-09	7.0E-09

GI-Tract Gastrointestinal Tract
St Stomach
SI Small Intestine
ULI Upper Large Intestine
LLI Lower Large Intestine

Table C-7.1. Biokinetic data for iodine

		Distribution (%)			Biological half-time (days)		
Age	f_1	Uptake by thyroid	Prompt urinary excretion	Faecal excretion from rest of body	Blood T_a	Thyroid T_b	Rest of body T_c
3 months	1	30	70	20	0.25	11.2	1.12
1 year	1	30	70	20	0.25	15	1.5
5 years	1	30	70	20	0.25	23	2.3
10 years	1	30	70	20	0.25	58	5.8
15 years	1	30	70	20	0.25	67	6.7
Adult	1	30	70	20	0.25	80	12

The biokinetic data summarized above refer to a recycling model.

Table C-7.2

Ingestion Dose Coefficients (Equivalent Dose (Sv/Bq) to Age 70 Years) for I-125 (T1/2 = 60.14 d)

Age at intake	3 Months	1 Year	5 Years	10 Years	15 Years	Adult
Adrenals	1.2E-10	8.6E-11	4.9E-11	3.9E-11	2.4E-11	2.7E-11
Bladder Wall	3.7E-10	3.0E-10	2.8E-10	2.2E-10	1.9E-10	1.5E-10
Bone Surfaces	5.5E-10	3.9E-10	2.5E-10	2.3E-10	1.6E-10	1.7E-10
Brain	1.3E-10	9.7E-11	5.9E-11	5.2E-11	3.4E-11	3.5E-11
Breast	1.3E-10	1.0E-10	4.9E-11	3.4E-11	2.1E-11	2.2E-11
GI-Tract						
St Wall	5.4E-10	3.2E-10	1.7E-10	1.0E-10	7.2E-11	6.3E-11
SI Wall	1.7E-10	1.2E-10	6.8E-11	4.6E-11	2.8E-11	2.8E-11
ULI Wall	4.9E-10	3.3E-10	1.6E-10	8.4E-11	4.8E-11	4.1E-11
LLI Wall	1.1E-09	7.2E-10	3.4E-10	1.6E-10	9.0E-11	7.3E-11
Kidneys	1.1E-10	8.1E-11	4.6E-11	3.7E-11	2.3E-11	2.6E-11
Liver	1.2E-10	8.9E-11	5.1E-11	4.0E-11	2.4E-11	2.6E-11
Lungs	2.9E-10	2.3E-10	1.3E-10	9.3E-11	4.2E-11	4.1E-11
Muscle	5.4E-10	4.5E-10	3.0E-10	2.8E-10	1.7E-10	1.4E-10
Ovaries	1.9E-10	1.4E-10	7.7E-11	5.1E-11	3.0E-11	3.0E-11
Pancreas	1.4E-10	1.0E-10	5.7E-11	4.5E-11	2.7E-11	2.9E-11
Red Marrow	1.3E-10	9.3E-11	5.6E-11	4.9E-11	3.4E-11	3.3E-11
Skin	1.3E-10	9.2E-11	5.2E-11	5.0E-11	3.4E-11	3.6E-11
Spleen	1.3E-10	9.3E-11	5.0E-11	4.0E-11	2.5E-11	2.7E-11
Testes	1.0E-10	7.3E-11	4.6E-11	3.4E-11	2.1E-11	2.3E-11
Thymus	1.7E-09	1.4E-09	5.1E-10	2.1E-10	6.7E-11	5.0E-11
Thyroid	1.0E-06	1.2E-06	8.4E-07	6.3E-07	4.4E-07	3.1E-07
Uterus	1.5E-10	1.1E-10	7.1E-11	5.0E-11	2.9E-11	3.0E-11
Remainder	4.2E-10	3.5E-10	2.4E-10	2.4E-10	1.5E-10	1.3E-10
Effective Dose	5.3E-08	5.8E-08	4.2E-08	3.2E-08	2.2E-08	1.5E-08

GI-Tract Gastrointestinal Tract
St Stomach
SI Small Intestine
ULI Upper Large Intestine
LLI Lower Large Intestine

Table C-7.3

Ingestion Dose Coefficients (Equivalent Dose (Sv/Bq) to Age 70 Years) for I-129 (T1/2 = 1.57E7 y)

Age at intake	3 Months	1 Year	5 Years	10 Years	15 Years	Adult
Adrenals	2.1E-10	1.6E-10	1.1E-10	1.3E-10	9.1E-11	1.2E-10
Bladder Wall	8.3E-10	6.9E-10	6.6E-10	6.0E-10	5.1E-10	4.3E-10
Bone Surfaces	5.5E-10	4.3E-10	3.1E-10	4.3E-10	3.4E-10	4.0E-10
Brain	2.3E-10	1.7E-10	1.2E-10	1.5E-10	1.1E-10	1.4E-10
Breast	2.3E-10	1.8E-10	1.1E-10	1.3E-10	8.7E-11	1.1E-10
GI-Tract						
St Wall	1.2E-09	7.2E-10	3.8E-10	2.8E-10	1.9E-10	2.0E-10
SI Wall	2.6E-10	2.0E-10	1.3E-10	1.4E-10	9.6E-11	1.3E-10
ULI Wall	1.4E-09	9.5E-10	5.0E-10	3.6E-10	2.2E-10	2.3E-10
LLI Wall	3.5E-09	2.4E-09	1.2E-09	7.8E-10	4.7E-10	4.3E-10
Kidneys	2.0E-10	1.6E-10	1.0E-10	1.3E-10	8.9E-11	1.2E-10
Liver	2.1E-10	1.7E-10	1.1E-10	1.3E-10	9.0E-11	1.2E-10
Lungs	3.5E-10	2.9E-10	1.9E-10	2.2E-10	1.3E-10	1.5E-10
Muscle	4.9E-10	4.3E-10	3.2E-10	4.4E-10	2.9E-10	3.0E-10
Ovaries	2.6E-10	2.0E-10	1.3E-10	1.4E-10	9.8E-11	1.3E-10
Pancreas	2.3E-10	1.7E-10	1.1E-10	1.3E-10	9.3E-11	1.2E-10
Red Marrow	2.3E-10	1.8E-10	1.2E-10	1.5E-10	1.1E-10	1.4E-10
Skin	2.3E-10	1.7E-10	1.1E-10	1.5E-10	1.1E-10	1.4E-10
Spleen	2.2E-10	1.7E-10	1.1E-10	1.3E-10	9.0E-11	1.2E-10
Testes	2.0E-10	1.5E-10	1.0E-10	1.2E-10	8.6E-11	1.2E-10
Thymus	1.4E-09	1.2E-09	5.7E-10	4.1E-10	1.7E-10	1.7E-10
Thyroid	3.7E-06	4.3E-06	3.4E-06	3.8E-06	2.8E-06	2.1E-06
Uterus	2.3E-10	1.8E-10	1.2E-10	1.4E-10	9.5E-11	1.3E-10
Remainder	4.1E-10	3.6E-10	2.8E-10	4.0E-10	2.8E-10	2.9E-10
Effective Dose	1.8E-07	2.2E-07	1.7E-07	1.9E-07	1.4E-07	1.1E-07

GI-Tract Gastrointestinal Tract
St Stomach
SI Small Intestine
ULI Upper Large Intestine
LLI Lower Large Intestine

Table C-7.4

Ingestion Dose Coefficients (Equivalent Dose (Sv/Bq) to Age 70 Years) for I-131 (T1/2 = 8.04 d)

Age at intake	3 Months	1 Year	5 Years	10 Years	15 Years	Adult
Adrenals	4.8E-10	3.4E-10	1.8E-10	1.1E-10	6.1E-11	5.1E-11
Bladder Wall	1.9E-09	1.5E-09	1.4E-09	1.2E-09	1.0E-09	7.6E-10
Bone Surfaces	6.2E-10	4.6E-10	2.9E-10	2.0E-10	1.5E-10	1.3E-10
Brain	5.3E-10	3.8E-10	2.5E-10	1.9E-10	1.5E-10	1.4E-10
Breast	5.7E-10	4.2E-10	2.3E-10	1.4E-10	7.3E-11	5.9E-11
GI-Tract						
St Wall	3.5E-09	2.0E-09	9.9E-10	5.7E-10	3.9E-10	3.1E-10
SI Wall	5.4E-10	3.6E-10	1.8E-10	1.1E-10	6.6E-11	5.4E-11
ULI Wall	1.7E-09	1.0E-09	4.5E-10	2.0E-10	1.1E-10	8.9E-11
LLI Wall	3.8E-09	2.2E-09	9.0E-10	3.8E-10	2.1E-10	1.6E-10
Kidneys	4.3E-10	2.9E-10	1.6E-10	8.9E-11	5.5E-11	4.6E-11
Liver	4.8E-10	3.3E-10	1.7E-10	1.0E-10	6.0E-11	4.9E-11
Lungs	7.3E-10	5.5E-10	3.3E-10	2.1E-10	1.3E-10	1.0E-10
Muscle	8.6E-10	6.6E-10	3.9E-10	2.5E-10	1.6E-10	1.3E-10
Ovaries	4.9E-10	3.3E-10	1.8E-10	1.0E-10	6.4E-11	5.2E-11
Pancreas	5.3E-10	3.7E-10	2.0E-10	1.2E-10	7.3E-11	6.1E-11
Red Marrow	5.1E-10	3.7E-10	2.2E-10	1.6E-10	1.2E-10	1.0E-10
Skin	4.8E-10	3.4E-10	1.8E-10	1.2E-10	8.1E-11	6.9E-11
Spleen	4.8E-10	3.4E-10	1.8E-10	1.1E-10	6.6E-11	5.4E-11
Testes	3.8E-10	2.5E-10	1.4E-10	8.1E-11	4.8E-11	4.0E-11
Thymus	2.3E-09	1.7E-09	8.5E-10	4.6E-10	2.2E-10	1.5E-10
Thyroid	3.7E-06	3.6E-06	2.1E-06	1.0E-06	6.8E-07	4.3E-07
Uterus	4.7E-10	3.2E-10	2.0E-10	1.2E-10	7.0E-11	5.9E-11
Remainder	7.6E-10	5.8E-10	3.5E-10	2.4E-10	1.6E-10	1.2E-10
Effective Dose	1.8E-07	1.8E-07	1.0E-07	5.2E-08	3.4E-08	2.2E-08

GI-Tract Gastrointestinal Tract
St Stomach
SI Small Intestine
ULI Upper Large Intestine
LLI Lower Large Intestine

Table C-7.5

Ingestion Dose Coefficients (Equivalent Dose (Sv/Bq) to Age 70 Years) for I-132 (T1/2 = 2.30 h)

Age at intake	3 Months	1 Year	5 Years	10 Years	15 Years	Adult
Adrenals	2.3E-10	1.6E-10	9.5E-11	6.3E-11	4.1E-11	3.6E-11
Bladder Wall	9.2E-10	7.2E-10	5.7E-10	4.8E-10	4.1E-10	3.1E-10
Bone Surfaces	2.0E-10	1.3E-10	7.2E-11	4.7E-11	3.0E-11	2.6E-11
Brain	1.6E-10	1.1E-10	5.9E-11	3.7E-11	2.3E-11	1.9E-11
Breast	1.8E-10	1.2E-10	6.3E-11	3.9E-11	2.5E-11	2.0E-11
GI-Tract						
St Wall	6.8E-09	3.9E-09	2.0E-09	1.2E-09	8.2E-10	6.5E-10
SI Wall	4.1E-10	2.8E-10	1.5E-10	9.7E-11	5.9E-11	4.8E-11
ULI Wall	4.4E-10	3.0E-10	1.7E-10	1.0E-10	6.5E-11	5.3E-11
LLI Wall	2.8E-10	1.9E-10	1.1E-10	7.2E-11	4.4E-11	3.8E-11
Kidneys	2.2E-10	1.5E-10	9.0E-11	5.9E-11	3.9E-11	3.3E-11
Liver	2.5E-10	1.7E-10	9.5E-11	5.8E-11	3.5E-11	2.9E-11
Lungs	2.0E-10	1.4E-10	7.6E-11	4.9E-11	3.2E-11	2.6E-11
Muscle	2.1E-10	1.4E-10	7.8E-11	5.0E-11	3.2E-11	2.7E-11
Ovaries	2.3E-10	1.6E-10	9.6E-11	6.3E-11	4.0E-11	3.3E-11
Pancreas	4.3E-10	3.1E-10	1.9E-10	1.3E-10	8.8E-11	7.4E-11
Red Marrow	1.7E-10	1.2E-10	6.9E-11	4.7E-11	3.1E-11	2.6E-11
Skin	1.6E-10	1.1E-10	5.6E-11	3.5E-11	2.2E-11	1.9E-11
Spleen	3.0E-10	2.1E-10	1.3E-10	8.8E-11	5.9E-11	5.3E-11
Testes	1.8E-10	1.2E-10	7.4E-11	4.7E-11	2.9E-11	2.4E-11
Thymus	2.3E-10	1.6E-10	8.0E-11	4.9E-11	3.0E-11	2.5E-11
Thyroid	4.0E-08	3.5E-08	1.9E-08	8.3E-09	5.4E-09	3.4E-09
Uterus	2.5E-10	1.8E-10	1.2E-10	7.5E-11	4.7E-11	3.9E-11
Remainder	2.0E-10	1.4E-10	7.7E-11	5.0E-11	3.3E-11	2.7E-11
Effective Dose	3.0E-09	2.4E-09	1.3E-09	6.2E-10	4.2E-10	2.9E-10

GI-Tract Gastrointestinal Tract
St Stomach
SI Small Intestine
ULI Upper Large Intestine
LLI Lower Large Intestine

Table C-8.1. Biokinetic data for caesium

		Distribution (%)		Biological half-time (days)	
		Total body		Total body	
Age	f_1	Comp A.	Comp. B	Comp. A	Comp. B
3 months	1	–	100	–	16
1 year	1	–	100	–	13
5 years	1	45	55	9.1	30
10 years	1	30	70	5.8	50
15 years	1	13	87	2.2	93
Adult	1	10	90	2	110

A urinary to faecal excretion ratio of 4:1 is assumed for caesium that has entered the transfer compartment.

Table C-8.2

Ingestion Dose Coefficients (Equivalent Dose (Sv/Bq) to Age 70 Years) for Cs-134 (T1/2 = 2.062 y)

Age at intake	3 Months	1 Year	5 Years	10 Years	15 Years	Adult
Adrenals	2.5E-08	1.6E-08	1.3E-08	1.5E-08	2.1E-08	2.1E-08
Bladder Wall	2.4E-08	1.4E-08	1.4E-08	1.4E-08	1.8E-08	2.0E-08
Bone Surfaces	2.4E-08	1.5E-08	1.3E-08	1.4E-08	1.9E-08	2.0E-08
Brain	2.2E-08	1.3E-08	1.1E-08	1.2E-08	1.5E-08	1.5E-08
Breast	1.8E-08	1.1E-08	9.0E-09	9.9E-09	1.4E-08	1.4E-08
GI-Tract						
St Wall	2.6E-08	1.5E-08	1.3E-08	1.3E-08	1.8E-08	1.9E-08
SI Wall	2.7E-08	1.7E-08	1.4E-08	1.6E-08	2.1E-08	2.1E-08
ULI Wall	3.2E-08	2.0E-08	1.6E-08	1.5E-08	2.0E-08	2.0E-08
LLI Wall	4.6E-08	2.9E-08	2.1E-08	1.9E-08	2.1E-08	2.3E-08
Kidneys	2.5E-08	1.5E-08	1.3E-08	1.4E-08	1.9E-08	2.0E-08
Liver	2.5E-08	1.5E-08	1.3E-08	1.4E-08	2.0E-08	2.0E-08
Lungs	2.2E-08	1.4E-08	1.1E-08	1.3E-08	1.8E-08	1.8E-08
Muscle	2.2E-08	1.3E-08	1.1E-08	1.3E-08	1.7E-08	1.7E-08
Ovaries	2.8E-08	1.7E-08	1.5E-08	1.6E-08	2.1E-08	2.2E-08
Pancreas	2.7E-08	1.6E-08	1.4E-08	1.6E-08	2.2E-08	2.2E-08
Red Marrow	2.1E-08	1.3E-08	1.1E-08	1.3E-08	1.8E-08	1.9E-08
Skin	1.8E-08	1.0E-08	8.6E-09	9.3E-09	1.3E-08	1.3E-08
Spleen	2.5E-08	1.5E-08	1.3E-08	1.4E-08	2.0E-08	2.0E-08
Testes	2.2E-08	1.3E-08	1.1E-08	1.2E-08	1.7E-08	1.7E-08
Thymus	2.4E-08	1.4E-08	1.2E-08	1.4E-08	1.9E-08	1.9E-08
Thyroid	2.6E-08	1.6E-08	1.3E-08	1.4E-08	1.9E-08	1.9E-08
Uterus	2.7E-08	1.6E-08	1.4E-08	1.6E-08	2.2E-08	2.2E-08
Remainder	2.2E-08	1.4E-08	1.1E-08	1.3E-08	1.9E-08	2.0E-08
Effective Dose	2.6E-08	1.6E-08	1.3E-08	1.4E-08	1.9E-08	1.9E-08

GI-Tract Gastrointestinal Tract
St Stomach
SI Small Intestine
ULI Upper Large Intestine
LLI Lower Large Intestine

Table C-8.3

Ingestion Dose Coefficients (Equivalent Dose (Sv/Bq) to Age 70 Years) for Cs-136 (T1/2 = 13.1 d)

Age at intake	3 Months	1 Year	5 Years	10 Years	15 Years	Adult
Adrenals	1.5E-08	9.8E-09	6.3E-09	4.7E-09	3.7E-09	3.3E-09
Bladder Wall	1.4E-08	8.9E-09	6.7E-09	4.3E-09	3.3E-09	3.2E-09
Bone Surfaces	1.4E-08	9.4E-09	6.0E-09	4.5E-09	3.6E-09	3.2E-09
Brain	1.2E-08	8.1E-09	5.1E-09	3.6E-09	2.7E-09	2.4E-09
Breast	9.9E-09	6.4E-09	4.0E-09	2.9E-09	2.5E-09	2.1E-09
GI-Tract						
St Wall	1.6E-08	1.0E-08	6.5E-09	4.4E-09	3.7E-09	3.3E-09
SI Wall	1.5E-08	1.0E-08	6.6E-09	4.9E-09	3.8E-09	3.3E-09
ULI Wall	1.8E-08	1.2E-08	7.2E-09	4.9E-09	3.7E-09	3.3E-09
LLI Wall	2.3E-08	1.6E-08	9.2E-09	5.9E-09	3.9E-09	3.7E-09
Kidneys	1.4E-08	9.3E-09	5.9E-09	4.4E-09	3.5E-09	3.1E-09
Liver	1.4E-08	9.4E-09	6.0E-09	4.4E-09	3.5E-09	3.1E-09
Lungs	1.3E-08	8.3E-09	5.2E-09	3.9E-09	3.2E-09	2.7E-09
Muscle	1.3E-08	8.3E-09	5.1E-09	3.8E-09	3.0E-09	2.7E-09
Ovaries	1.6E-08	1.1E-08	6.8E-09	5.0E-09	3.9E-09	3.4E-09
Pancreas	1.6E-08	1.0E-08	6.6E-09	5.0E-09	3.9E-09	3.5E-09
Red Marrow	1.2E-08	7.8E-09	5.2E-09	4.0E-09	3.2E-09	2.9E-09
Skin	9.6E-09	6.2E-09	3.8E-09	2.7E-09	2.1E-09	1.9E-09
Spleen	1.4E-08	9.3E-09	5.9E-09	4.4E-09	3.5E-09	3.1E-09
Testes	1.2E-08	7.9E-09	5.1E-09	3.7E-09	3.0E-09	2.7E-09
Thymus	1.3E-08	8.7E-09	5.5E-09	4.1E-09	3.3E-09	2.9E-09
Thyroid	1.5E-08	9.7E-09	6.1E-09	4.4E-09	3.3E-09	2.9E-09
Uterus	1.5E-08	1.0E-08	6.7E-09	5.0E-09	3.9E-09	3.5E-09
Remainder	1.3E-08	8.3E-09	5.2E-09	3.8E-09	3.5E-09	3.1E-09
Effective Dose	1.5E-08	9.6E-09	6.1E-09	4.4E-09	3.5E-09	3.1E-09

GI-Tract Gastrointestinal Tract
St Stomach
SI Small Intestine
ULI Upper Large Intestine
LLI Lower Large Intestine

Table C-8.4

Ingestion Dose Coefficients (Equivalent Dose (Sv/Bq) to Age 70 Years) for Cs-137 (T1/2 = 30.0 y)

Age at intake	3 Months	1 Year	5 Years	10 Years	15 Years	Adult
Adrenals	1.9E-08	1.1E-08	9.1E-09	1.0E-08	1.4E-08	1.4E-08
Bladder Wall	2.0E-08	1.2E-08	1.1E-08	1.1E-08	1.4E-08	1.4E-08
Bone Surfaces	1.9E-08	1.1E-08	8.8E-09	9.8E-09	1.3E-08	1.4E-08
Brain	1.8E-08	1.0E-08	8.2E-09	8.9E-09	1.2E-08	1.2E-08
Breast	1.6E-08	9.2E-09	7.4E-09	8.1E-09	1.1E-08	1.1E-08
GI-Tract						
St Wall	2.2E-08	1.2E-08	9.5E-09	9.8E-09	1.3E-08	1.3E-08
SI Wall	1.9E-08	1.1E-08	9.4E-09	1.0E-08	1.4E-08	1.4E-08
ULI Wall	2.9E-08	1.8E-08	1.3E-08	1.2E-08	1.4E-08	1.4E-08
LLI Wall	4.8E-08	3.1E-08	1.9E-08	1.6E-08	1.6E-08	1.7E-08
Kidneys	1.9E-08	1.1E-08	8.8E-09	9.8E-09	1.3E-08	1.3E-08
Liver	1.9E-08	1.1E-08	8.9E-09	9.8E-09	1.3E-08	1.4E-08
Lungs	1.8E-08	1.0E-08	8.4E-09	9.2E-09	1.3E-08	1.3E-08
Muscle	1.8E-08	1.0E-08	8.3E-09	9.1E-09	1.2E-08	1.3E-08
Ovaries	2.0E-08	1.2E-08	9.5E-09	1.0E-08	1.4E-08	1.4E-08
Pancreas	1.9E-08	1.1E-08	9.3E-09	1.0E-08	1.4E-08	1.4E-08
Red Marrow	1.7E-08	1.0E-08	8.3E-09	9.4E-09	1.3E-08	1.3E-08
Skin	1.6E-08	9.0E-09	7.3E-09	7.8E-09	1.1E-08	1.1E-08
Spleen	1.9E-08	1.1E-08	8.8E-09	9.8E-09	1.3E-08	1.4E-08
Testes	1.7E-08	1.0E-08	8.2E-09	9.0E-09	1.2E-08	1.3E-08
Thymus	1.8E-08	1.0E-08	8.6E-09	9.5E-09	1.3E-08	1.3E-08
Thyroid	1.9E-08	1.1E-08	9.0E-09	9.8E-09	1.3E-08	1.3E-08
Uterus	1.9E-08	1.1E-08	9.4E-09	1.0E-08	1.4E-08	1.4E-08
Remainder	1.8E-08	1.0E-08	8.3E-09	9.1E-09	1.2E-08	1.3E-08
Effective Dose	2.1E-08	1.2E-08	9.7E-09	1.0E-08	1.3E-08	1.4E-08

GI-Tract Gastrointestinal Tract
St Stomach
SI Small Intestine
ULI Upper Large Intestine
LLI Lower Large Intestine

Table C-9.1. Biokinetic data for cerium

		Distribution (%)			Biological half-time (days)		
Age	f_1	Skeleton	Liver	Other tissues	Skeleton	Liver	Other tissues
3 months	$5 \cdot 10^{-3}$	70	10	20	3500	3500	3500
1 year	$5 \cdot 10^{-4}$	50	30	20	3500	3500	3500
5 years	$5 \cdot 10^{-4}$	50	30	20	3500	3500	3500
10 years	$5 \cdot 10^{-4}$	40	40	20	3500	3500	3500
15 years	$5 \cdot 10^{-4}$	40	40	20	3500	3500	3500
Adult	$5 \cdot 10^{-4}$	30	50	20	3500	3500	3500

A urinary to faecal excretion ratio of 1:9 is assumed for cerium that has entered the transfer compartment.

Table C-9.2

Ingestion Dose Coefficients (Equivalent Dose (Sv/Bq) to Age 70 Years) for Ce-141 (T1/2 = 32.501 d)

Age at intake	3 Months	1 Year	5 Years	10 Years	15 Years	Adult
Adrenals	7.8E-11	3.9E-11	2.1E-11	1.2E-11	7.3E-12	5.4E-12
Bladder Wall	2.3E-10	1.6E-10	9.3E-11	6.2E-11	4.0E-11	3.2E-11
Bone Surfaces	4.5E-09	3.4E-10	2.2E-10	1.1E-10	6.5E-11	4.9E-11
Brain	2.5E-11	1.9E-12	9.4E-13	5.3E-13	3.2E-13	2.5E-13
Breast	3.6E-11	1.0E-11	5.2E-12	2.6E-12	1.1E-12	7.6E-13
GI-Tract						
St Wall	2.7E-09	1.5E-09	7.4E-10	4.2E-10	2.9E-10	2.3E-10
SI Wall	6.3E-09	4.1E-09	2.1E-09	1.3E-09	7.2E-10	5.8E-10
ULI Wall	3.5E-08	2.3E-08	1.2E-08	6.9E-09	3.8E-09	3.1E-09
LLI Wall	9.8E-08	6.3E-08	3.2E-08	1.9E-08	1.1E-08	8.7E-09
Kidneys	1.0E-10	5.9E-11	3.6E-11	2.3E-11	1.5E-11	1.2E-11
Liver	3.6E-10	1.2E-10	6.4E-11	4.6E-11	2.7E-11	2.4E-11
Lungs	4.5E-11	1.5E-11	7.2E-12	4.0E-12	2.2E-12	1.4E-12
Muscle	9.5E-11	5.4E-11	3.1E-11	2.0E-11	1.4E-11	1.1E-11
Ovaries	6.4E-10	4.7E-10	2.9E-10	2.0E-10	1.3E-10	1.0E-10
Pancreas	1.3E-10	7.6E-11	4.1E-11	2.6E-11	1.6E-11	1.1E-11
Red Marrow	1.1E-09	8.7E-11	5.5E-11	3.6E-11	2.6E-11	1.9E-11
Skin	4.8E-11	1.9E-11	1.0E-11	6.2E-12	4.0E-12	3.2E-12
Spleen	9.4E-11	5.4E-11	3.0E-11	1.9E-11	1.1E-11	8.9E-12
Testes	1.1E-10	6.3E-11	3.3E-11	2.0E-11	1.1E-11	7.9E-12
Thymus	3.2E-11	7.7E-12	3.3E-12	1.6E-12	9.1E-13	6.3E-13
Thyroid	2.8E-11	4.4E-12	1.7E-12	8.3E-13	4.1E-13	2.9E-13
Uterus	3.5E-10	2.5E-10	1.5E-10	9.6E-11	6.0E-11	4.5E-11
Remainder	2.3E-10	1.4E-10	7.8E-11	4.6E-11	2.8E-11	2.2E-11
Effective Dose	8.1E-09	5.1E-09	2.6E-09	1.5E-09	8.9E-10	7.1E-10

GI-Tract Gastrointestinal Tract
St Stomach
SI Small Intestine
ULI Upper Large Intestine
LLI Lower Large Intestine

Table C-9.3

Ingestion Dose Coefficients (Equivalent Dose (Sv/Bq) to Age 70 Years) for Ce-144 (T1/2 = 284.3 d)

Age at intake	3 Months	1 Year	5 Years	10 Years	15 Years	Adult
Adrenals	9.0E-10	9.1E-11	5.1E-11	3.1E-11	1.9E-11	1.6E-11
Bladder Wall	9.8E-10	1.7E-10	9.4E-11	6.1E-11	3.7E-11	3.0E-11
Bone Surfaces	3.5E-08	2.3E-09	1.6E-09	7.0E-10	4.4E-10	3.3E-10
Brain	8.5E-10	6.7E-11	3.7E-11	2.2E-11	1.4E-11	1.2E-11
Breast	8.5E-10	7.3E-11	4.0E-11	2.4E-11	1.4E-11	1.2E-11
GI-Tract						
St Wall	1.4E-08	7.5E-09	3.6E-09	2.0E-09	1.4E-09	1.1E-09
SI Wall	4.4E-08	2.8E-08	1.4E-08	8.3E-09	4.6E-09	3.7E-09
ULI Wall	2.7E-07	1.7E-07	8.7E-08	5.1E-08	2.9E-08	2.3E-08
LLI Wall	7.6E-07	4.9E-07	2.5E-07	1.5E-07	8.3E-08	6.7E-08
Kidneys	9.1E-10	1.0E-10	6.0E-11	3.7E-11	2.3E-11	2.0E-11
Liver	1.2E-08	2.9E-09	1.7E-09	1.4E-09	9.4E-10	9.6E-10
Lungs	8.7E-10	7.7E-11	4.2E-11	2.5E-11	1.6E-11	1.3E-11
Muscle	9.0E-10	1.0E-10	5.7E-11	3.5E-11	2.2E-11	1.8E-11
Ovaries	1.3E-09	3.8E-10	2.2E-10	1.5E-10	9.7E-11	7.5E-11
Pancreas	9.2E-10	1.1E-10	6.2E-11	3.8E-11	2.4E-11	1.9E-11
Red Marrow	4.4E-08	2.4E-09	1.3E-09	5.5E-10	3.3E-10	1.9E-10
Skin	8.6E-10	7.9E-11	4.3E-11	2.6E-11	1.6E-11	1.4E-11
Spleen	9.0E-10	9.9E-11	5.5E-11	3.3E-11	2.0E-11	1.7E-11
Testes	9.0E-10	1.1E-10	5.8E-11	3.5E-11	2.1E-11	1.7E-11
Thymus	8.5E-10	7.0E-11	3.9E-11	2.3E-11	1.4E-11	1.2E-11
Thyroid	8.5E-10	6.8E-11	3.7E-11	2.2E-11	1.4E-11	1.2E-11
Uterus	1.1E-09	2.3E-10	1.2E-10	8.0E-11	4.9E-11	3.8E-11
Remainder	2.0E-09	7.5E-10	4.1E-10	2.2E-10	1.2E-10	9.5E-11
Effective Dose	6.6E-08	3.9E-08	1.9E-08	1.1E-08	6.5E-09	5.2E-09

GI-Tract Gastrointestinal Tract
St Stomach
SI Small Intestine
ULI Upper Large Intestine
LLI Lower Large Intestine

Table C-10.1

Ingestion Dose Coefficients (Equivalent Dose (Sv/Bq) to Age 70 Years) for Pu-238 (T1/2 = 87.74 y)*

Age at intake	3 Months	1 Year	5 Years	10 Years	15 Years	Adult
Adrenals	5.1E-07	4.4E-08	2.9E-08	2.0E-08	1.6E-08	1.3E-08
Bladder Wall	5.1E-07	4.4E-08	2.9E-08	2.0E-08	1.6E-08	1.3E-08
Bone Surfaces	6.9E-05	7.0E-06	6.3E-06	6.1E-06	6.5E-06	7.5E-06
Brain	5.1E-07	4.4E-08	2.9E-08	2.0E-08	1.6E-08	1.3E-08
Breast	5.1E-07	4.4E-08	2.9E-08	2.0E-08	1.6E-08	1.3E-08
GI-Tract						
St Wall	5.3E-07	5.3E-08	3.3E-08	2.3E-08	1.7E-08	1.4E-08
SI Wall	5.5E-07	6.9E-08	4.1E-08	2.7E-08	2.0E-08	1.6E-08
ULI Wall	7.4E-07	1.9E-07	1.0E-07	6.4E-08	4.0E-08	3.2E-08
LLI Wall	1.2E-06	4.6E-07	2.4E-07	1.4E-07	8.6E-08	7.0E-08
Kidneys	1.5E-06	1.2E-07	7.3E-08	5.0E-08	3.8E-08	3.2E-08
Liver	2.5E-05	2.4E-06	2.2E-06	1.8E-06	1.5E-06	1.6E-06
Lungs	5.1E-07	4.4E-08	2.9E-08	2.0E-08	1.6E-08	1.3E-08
Muscle	5.1E-07	4.4E-08	2.9E-08	2.0E-08	1.6E-08	1.3E-08
Ovaries	1.4E-06	1.4E-07	1.4E-07	1.3E-07	1.2E-07	9.7E-08
Pancreas	5.1E-07	4.4E-08	2.9E-08	2.0E-08	1.6E-08	1.3E-08
Red Marrow	1.1E-05	9.6E-07	5.7E-07	4.1E-07	3.5E-07	3.7E-07
Skin	5.1E-07	4.4E-08	2.9E-08	2.0E-08	1.6E-08	1.3E-08
Spleen	5.1E-07	4.4E-08	2.9E-08	2.0E-08	1.6E-08	1.3E-08
Testes	1.6E-06	1.7E-07	1.3E-07	1.1E-07	1.2E-07	9.9E-08
Thymus	5.1E-07	4.4E-08	2.9E-08	2.0E-08	1.6E-08	1.3E-08
Thyroid	5.1E-07	4.4E-08	2.9E-08	2.0E-08	1.6E-08	1.3E-08
Uterus	5.1E-07	4.4E-08	2.9E-08	2.0E-08	1.6E-08	1.3E-08
Remainder	5.3E-07	4.6E-08	3.0E-08	2.1E-08	1.6E-08	1.3E-08
Effective Dose	4.0E-06	4.0E-07	3.1E-07	2.5E-07	2.2E-07	2.3E-07

GI-Tract Gastrointestinal Tract
St Stomach
SI Small Intestine
ULI Upper Large Intestine
LLI Lower Large Intestine

* In the biokinetic model for Pu parameter values for the adult apply to ages > 25 y. For radioisotopes of this element the dose coefficients for the adult are based on the 50-y integrated doses following an acute intake at age 25 y.

Table C-10.2

Ingestion Dose Coefficients (Equivalent Dose (Sv/Bq) to Age 70 Years) for Pu-239 (T1/2 = 24065 y)*

Age at intake	3 Months	1 Year	5 Years	10 Years	15 Years	Adult
Adrenals	5.5E-07	4.8E-08	3.2E-08	2.3E-08	1.8E-08	1.5E-08
Bladder Wall	5.5E-07	4.8E-08	3.2E-08	2.3E-08	1.8E-08	1.5E-08
Bone Surfaces	7.6E-05	7.7E-06	7.2E-06	6.9E-06	7.4E-06	8.4E-06
Brain	5.5E-07	4.8E-08	3.2E-08	2.3E-08	1.8E-08	1.5E-08
Breast	5.5E-07	4.8E-08	3.2E-08	2.3E-08	1.8E-08	1.5E-08
GI-Tract						
St Wall	5.6E-07	5.6E-08	3.6E-08	2.5E-08	1.9E-08	1.6E-08
SI Wall	5.8E-07	7.1E-08	4.3E-08	3.0E-08	2.1E-08	1.8E-08
ULI Wall	7.6E-07	1.9E-07	1.0E-07	6.4E-08	4.0E-08	3.3E-08
LLI Wall	1.2E-06	4.4E-07	2.3E-07	1.4E-07	8.4E-08	6.8E-08
Kidneys	1.4E-06	1.2E-07	7.5E-08	5.2E-08	4.0E-08	3.4E-08
Liver	2.7E-05	2.6E-06	2.5E-06	2.0E-06	1.7E-06	1.7E-06
Lungs	5.5E-07	4.8E-08	3.2E-08	2.3E-08	1.8E-08	1.5E-08
Muscle	5.5E-07	4.8E-08	3.2E-08	2.3E-08	1.8E-08	1.5E-08
Ovaries	1.5E-06	1.6E-07	1.6E-07	1.4E-07	1.3E-07	1.1E-07
Pancreas	5.5E-07	4.8E-08	3.2E-08	2.3E-08	1.8E-08	1.5E-08
Red Marrow	1.1E-05	9.6E-07	5.9E-07	4.4E-07	3.8E-07	4.0E-07
Skin	5.5E-07	4.8E-08	3.2E-08	2.3E-08	1.8E-08	1.5E-08
Spleen	5.5E-07	4.8E-08	3.2E-08	2.3E-08	1.8E-08	1.5E-08
Testes	1.7E-06	1.8E-07	1.4E-07	1.2E-07	1.3E-07	1.1E-07
Thymus	5.5E-07	4.8E-08	3.2E-08	2.3E-08	1.8E-08	1.5E-08
Thyroid	5.5E-07	4.8E-08	3.2E-08	2.3E-08	1.8E-08	1.5E-08
Uterus	5.5E-07	4.8E-08	3.2E-08	2.3E-08	1.8E-08	1.5E-08
Remainder	5.7E-07	5.0E-08	3.3E-08	2.3E-08	1.8E-08	1.5E-08
Effective Dose	4.2E-06	4.2E-07	3.3E-07	2.7E-07	2.5E-07	2.5E-07

GI-Tract Gastrointestinal Tract
St Stomach
SI Small Intestine
ULI Upper Large Intestine
LLI Lower Large Intestine

* In the biokinetic model for Pu parameter values for the adult apply to ages > 25 y. For radioisotopes of this element the dose coefficients for the adult are based on the 50-y integrated doses following an acute intake at age 25 y.

Table C-10.3

Ingestion Dose Coefficients (Equivalent Dose (Sv/Bq) to Age 70 Years) for Pu-240 (T1/2 = 6537 y)*

Age at intake	3 Months	1 Year	5 Years	10 Years	15 Years	Adult
Adrenals	5.5E-07	4.8E-08	3.2E-08	2.3E-08	1.8E-08	1.5E-08
Bladder Wall	5.5E-07	4.8E-08	3.2E-08	2.3E-08	1.8E-08	1.5E-08
Bone Surfaces	7.6E-05	7.7E-06	7.2E-06	6.9E-06	7.4E-06	8.4E-06
Brain	5.5E-07	4.8E-08	3.2E-08	2.3E-08	1.8E-08	1.5E-08
Breast	5.5E-07	4.8E-08	3.2E-08	2.3E-08	1.8E-08	1.5E-08
GI-Tract						
St Wall	5.7E-07	5.7E-08	3.6E-08	2.5E-08	1.9E-08	1.6E-08
SI Wall	5.9E-07	7.1E-08	4.3E-08	3.0E-08	2.1E-08	1.8E-08
ULI Wall	7.6E-07	1.9E-07	1.0E-07	6.4E-08	4.1E-08	3.3E-08
LLI Wall	1.2E-06	4.4E-07	2.3E-07	1.4E-07	8.4E-08	6.8E-08
Kidneys	1.4E-06	1.2E-07	7.5E-08	5.2E-08	4.0E-08	3.4E-08
Liver	2.7E-05	2.6E-06	2.5E-06	2.0E-06	1.7E-06	1.7E-06
Lungs	5.5E-07	4.8E-08	3.2E-08	2.3E-08	1.8E-08	1.5E-08
Muscle	5.5E-07	4.8E-08	3.2E-08	2.3E-08	1.8E-08	1.5E-08
Ovaries	1.5E-06	1.6E-07	1.6E-07	1.4E-07	1.3E-07	1.1E-07
Pancreas	5.5E-07	4.8E-08	3.2E-08	2.3E-08	1.8E-08	1.5E-08
Red Marrow	1.1E-05	9.6E-07	5.9E-07	4.4E-07	3.8E-07	4.0E-07
Skin	5.5E-07	4.8E-08	3.2E-08	2.3E-08	1.8E-08	1.5E-08
Spleen	5.5E-07	4.8E-08	3.2E-08	2.3E-08	1.8E-08	1.5E-08
Testes	1.7E-06	1.8E-07	1.4E-07	1.2E-07	1.3E-07	1.1E-07
Thymus	5.5E-07	4.8E-08	3.2E-08	2.3E-08	1.8E-08	1.5E-08
Thyroid	5.5E-07	4.8E-08	3.2E-08	2.3E-08	1.8E-08	1.5E-08
Uterus	5.5E-07	4.8E-08	3.2E-08	2.3E-08	1.8E-08	1.5E-08
Remainder	5.7E-07	5.0E-08	3.3E-08	2.3E-08	1.8E-08	1.5E-08
Effective Dose	4.2E-06	4.2E-07	3.3E-07	2.7E-07	2.5E-07	2.5E-07

GI-Tract Gastrointestinal Tract
St Stomach
SI Small Intestine
ULI Upper Large Intestine
LLI Lower Large Intestine

* In the biokinetic model for Pu parameter values for the adult apply to ages > 25 y. For radioisotopes of this element the dose coefficients for the adult are based on the 50-y integrated doses following an acute intake at age 25 y.

Table C-10.4

Ingestion Dose Coefficients (Equivalent Dose (Sv/Bq) to Age 70 Years) for Pu-241 (T1/2 = 14.4 y)*

Age at intake	3 Months	1 Year	5 Years	10 Years	15 Years	Adult
Adrenals	8.1E-09	7.6E-10	5.8E-10	4.6E-10	3.7E-10	3.0E-10
Bladder Wall	8.1E-09	7.6E-10	5.8E-10	4.6E-10	3.7E-10	3.0E-10
Bone Surfaces	1.2E-06	1.3E-07	1.4E-07	1.4E-07	1.5E-07	1.7E-07
Brain	8.1E-09	7.6E-10	5.8E-10	4.6E-10	3.7E-10	3.0E-10
Breast	8.1E-09	7.6E-10	5.8E-10	4.6E-10	3.7E-10	3.0E-10
GI-Tract						
St Wall	8.1E-09	8.0E-10	6.0E-10	4.7E-10	3.7E-10	3.0E-10
SI Wall	8.2E-09	8.7E-10	6.4E-10	4.9E-10	3.9E-10	3.1E-10
ULI Wall	9.1E-09	1.5E-09	9.3E-10	6.7E-10	4.8E-10	3.9E-10
LLI Wall	1.1E-08	2.8E-09	1.6E-09	1.0E-09	7.0E-10	5.7E-10
Kidneys	1.0E-08	9.8E-10	7.8E-10	6.4E-10	5.5E-10	5.0E-10
Liver	4.6E-07	4.6E-08	4.5E-08	3.9E-08	3.5E-08	3.4E-08
Lungs	8.1E-09	7.6E-10	5.8E-10	4.6E-10	3.7E-10	3.0E-10
Muscle	8.1E-09	7.6E-10	5.8E-10	4.6E-10	3.7E-10	3.0E-10
Ovaries	2.7E-08	2.8E-09	2.8E-09	2.7E-09	2.6E-09	2.2E-09
Pancreas	8.1E-09	7.6E-10	5.8E-10	4.6E-10	3.7E-10	3.0E-10
Red Marrow	8.5E-08	8.4E-09	7.4E-09	6.8E-09	6.4E-09	6.5E-09
Skin	8.1E-09	7.6E-10	5.8E-10	4.6E-10	3.7E-10	3.0E-10
Spleen	8.1E-09	7.6E-10	5.8E-10	4.6E-10	3.7E-10	3.0E-10
Testes	2.7E-08	2.8E-09	2.5E-09	2.3E-09	2.5E-09	2.2E-09
Thymus	8.1E-09	7.6E-10	5.8E-10	4.6E-10	3.7E-10	3.0E-10
Thyroid	8.1E-09	7.6E-10	5.8E-10	4.6E-10	3.7E-10	3.0E-10
Uterus	8.1E-09	7.6E-10	5.8E-10	4.6E-10	3.7E-10	3.0E-10
Remainder	8.1E-09	7.6E-10	5.8E-10	4.6E-10	3.7E-10	3.0E-10
Effective Dose	5.7E-08	5.8E-09	5.5E-09	5.1E-09	4.8E-09	4.8E-09

GI-Tract	Gastrointestinal Tract
St	Stomach
SI	Small Intestine
ULI	Upper Large Intestine
LLI	Lower Large Intestine

* In the biokinetic model for Pu parameter values for the adult apply to ages > 25 y. For radioisotopes of this element the dose coefficients for the adult are based on the 50-y integrated doses following an acute intake at age 25 y.

Table C-11.1

Ingestion Dose Coefficients (Equivalent Dose (Sv/Bq) to Age 70 Years) for Am-241 (T1/2 = 432.2 y)*

Age at intake	3 Months	1 Year	5 Years	10 Years	15 Years	Adult
Adrenals	4.2E-07	3.8E-08	2.9E-08	2.2E-08	1.8E-08	1.5E-08
Bladder Wall	4.2E-07	3.8E-08	2.9E-08	2.2E-08	1.8E-08	1.5E-08
Bone Surfaces	8.4E-05	8.5E-06	7.5E-06	7.5E-06	8.2E-06	9.2E-06
Brain	4.2E-07	3.8E-08	2.9E-08	2.2E-08	1.8E-08	1.5E-08
Breast	4.2E-07	3.8E-08	2.9E-08	2.2E-08	1.8E-08	1.5E-08
GI-Tract						
St Wall	4.4E-07	4.8E-08	3.3E-08	2.5E-08	1.9E-08	1.7E-08
SI Wall	4.6E-07	6.4E-08	4.2E-08	3.0E-08	2.2E-08	1.9E-08
ULI Wall	6.6E-07	1.9E-07	1.1E-07	6.8E-08	4.3E-08	3.6E-08
LLI Wall	1.1E-06	4.7E-07	2.5E-07	1.5E-07	9.1E-08	7.5E-08
Kidneys	1.7E-06	1.4E-07	9.5E-08	6.6E-08	5.1E-08	4.6E-08
Liver	9.1E-06	8.7E-07	8.4E-07	6.0E-07	5.1E-07	5.5E-07
Lungs	4.2E-07	3.8E-08	2.9E-08	2.2E-08	1.8E-08	1.5E-08
Muscle	4.2E-07	3.8E-08	2.9E-08	2.2E-08	1.8E-08	1.5E-08
Ovaries	2.2E-06	2.2E-07	2.3E-07	2.1E-07	2.0E-07	1.8E-07
Pancreas	4.2E-07	3.8E-08	2.9E-08	2.2E-08	1.8E-08	1.5E-08
Red Marrow	1.3E-05	1.1E-06	6.3E-07	4.3E-07	3.5E-07	3.1E-07
Skin	4.2E-07	3.8E-08	2.9E-08	2.2E-08	1.8E-08	1.5E-08
Spleen	4.2E-07	3.8E-08	2.9E-08	2.2E-08	1.8E-08	1.5E-08
Testes	2.4E-06	2.5E-07	2.1E-07	1.9E-07	1.9E-07	1.7E-07
Thymus	4.2E-07	3.8E-08	2.9E-08	2.2E-08	1.8E-08	1.5E-08
Thyroid	4.2E-07	3.8E-08	2.9E-08	2.2E-08	1.8E-08	1.5E-08
Uterus	4.2E-07	3.8E-08	2.9E-08	2.2E-08	1.8E-08	1.5E-08
Remainder	4.4E-07	4.1E-08	3.0E-08	2.3E-08	1.8E-08	1.6E-08
Effective Dose	3.7E-06	3.8E-07	2.8E-07	2.2E-07	2.1E-07	2.1E-07

GI-Tract	Gastrointestinal Tract
St	Stomach
SI	Small Intestine
ULI	Upper Large Intestine
LLI	Lower Large Intestine

* In the biokinetic model for Am parameter values for the adult apply to ages > 25 y. For radioisotopes of this element the dose coefficients for the adult are based on the 50-y integrated doses following an acute intake at age 25 y.

Table C-12.1

Ingestion Dose Coefficients (Equivalent Dose (Sv/Bq) to Age 70 Years) for Np-237 (T1/2 = 2.14E6 y)*

Age at intake	3 Months	1 Year	5 Years	10 Years	15 Years	Adult
Adrenals	2.9E-07	2.6E-08	1.7E-08	1.2E-08	9.6E-09	7.2E-09
Bladder Wall	2.9E-07	2.6E-08	1.7E-08	1.2E-08	9.6E-09	7.2E-09
Bone Surfaces	5.0E-05	5.1E-06	4.1E-06	4.2E-06	4.9E-06	5.5E-06
Brain	2.9E-07	2.6E-08	1.7E-08	1.2E-08	9.6E-09	7.2E-09
Breast	2.9E-07	2.6E-08	1.7E-08	1.2E-08	9.5E-09	7.1E-09
GI-Tract						
St Wall	3.1E-07	3.4E-08	2.1E-08	1.5E-08	1.1E-08	8.4E-09
SI Wall	3.3E-07	4.8E-08	2.8E-08	1.9E-08	1.3E-08	1.0E-08
ULI Wall	5.1E-07	1.6E-07	8.6E-08	5.3E-08	3.2E-08	2.5E-08
LLI Wall	9.0E-07	4.2E-07	2.1E-07	1.3E-07	7.6E-08	6.0E-08
Kidneys	1.2E-06	9.5E-08	5.3E-08	3.5E-08	2.5E-08	1.9E-08
Liver	2.1E-06	1.9E-07	1.4E-07	9.6E-08	7.8E-08	8.7E-08
Lungs	2.9E-07	2.6E-08	1.7E-08	1.2E-08	9.6E-09	7.2E-09
Muscle	2.9E-07	2.6E-08	1.7E-08	1.2E-08	9.6E-09	7.2E-09
Ovaries	1.1E-06	1.2E-07	1.1E-07	1.0E-07	9.5E-08	7.2E-08
Pancreas	2.9E-07	2.6E-08	1.7E-08	1.2E-08	9.6E-09	7.2E-09
Red Marrow	7.3E-06	6.4E-07	3.9E-07	2.6E-07	2.1E-07	2.1E-07
Skin	2.9E-07	2.6E-08	1.7E-08	1.2E-08	9.6E-09	7.1E-09
Spleen	2.9E-07	2.6E-08	1.7E-08	1.2E-08	9.6E-09	7.2E-09
Testes	1.4E-06	1.4E-07	9.9E-08	8.4E-08	9.4E-08	7.4E-08
Thymus	2.9E-07	2.6E-08	1.7E-08	1.2E-08	9.6E-09	7.1E-09
Thyroid	2.9E-07	2.6E-08	1.7E-08	1.2E-08	9.6E-09	7.2E-09
Uterus	2.9E-07	2.6E-08	1.7E-08	1.2E-08	9.6E-09	7.2E-09
Remainder	3.1E-07	2.7E-08	1.8E-08	1.3E-08	9.8E-09	7.3E-09
Effective Dose	2.0E-06	2.1E-07	1.4E-07	1.2E-07	1.1E-07	1.1E-07

GI-Tract	Gastrointestinal Tract
St	Stomach
SI	Small Intestine
ULI	Upper Large Intestine
LLI	Lower Large Intestine

* In the biokinetic model for Np parameter values for the adult apply to ages > 25 y. For radioisotopes of this element the dose coefficients for the adult are based on the 50-y integrated doses following an acute intake at age 25 y.

Table C-12.2

Ingestion Dose Coefficients (Equivalent Dose (Sv/Bq) to Age 70 Years) for Np-239 (T1/2 = 2.355 d)*

Age at intake	3 Months	1 Year	5 Years	10 Years	15 Years	Adult
Adrenals	8.6E-11	6.2E-11	3.5E-11	2.1E-11	1.2E-11	9.0E-12
Bladder Wall	3.0E-10	2.2E-10	1.3E-10	9.0E-11	5.8E-11	4.7E-11
Bone Surfaces	4.7E-10	1.3E-10	7.1E-11	4.5E-11	3.0E-11	2.4E-11
Brain	5.1E-12	7.5E-13	3.3E-13	1.5E-13	7.0E-14	4.9E-14
Breast	2.7E-11	1.6E-11	8.1E-12	4.0E-12	1.6E-12	1.1E-12
GI-Tract						
St Wall	4.2E-09	2.3E-09	1.1E-09	6.6E-10	4.5E-10	3.5E-10
SI Wall	9.2E-09	6.0E-09	3.1E-09	1.9E-09	1.1E-09	8.6E-10
ULI Wall	4.5E-08	3.0E-08	1.5E-08	8.8E-09	4.9E-09	3.9E-09
LLI Wall	9.8E-08	6.4E-08	3.2E-08	1.9E-08	1.1E-08	8.7E-09
Kidneys	1.5E-10	9.4E-11	6.0E-11	3.9E-11	2.6E-11	2.1E-11
Liver	1.6E-10	1.1E-10	5.9E-11	3.5E-11	1.8E-11	1.4E-11
Lungs	3.7E-11	2.3E-11	1.1E-11	6.2E-12	3.4E-12	2.1E-12
Muscle	1.2E-10	8.2E-11	4.7E-11	3.1E-11	2.1E-11	1.7E-11
Ovaries	1.0E-09	7.6E-10	4.4E-10	3.0E-10	2.0E-10	1.5E-10
Pancreas	1.6E-10	1.2E-10	6.7E-11	4.3E-11	2.7E-11	2.0E-11
Red Marrow	1.4E-10	6.7E-11	5.3E-11	4.3E-11	3.2E-11	2.6E-11
Skin	4.4E-11	2.8E-11	1.5E-11	9.4E-12	6.1E-12	4.9E-12
Spleen	1.2E-10	8.4E-11	4.8E-11	3.1E-11	1.9E-11	1.5E-11
Testes	1.3E-10	8.9E-11	4.7E-11	3.0E-11	1.6E-11	1.2E-11
Thymus	1.8E-11	1.1E-11	4.9E-12	2.2E-12	1.2E-12	8.0E-13
Thyroid	1.3E-11	5.4E-12	1.7E-12	7.6E-13	2.6E-13	1.4E-13
Uterus	4.9E-10	3.6E-10	2.2E-10	1.5E-10	9.2E-11	7.0E-11
Remainder	3.2E-10	2.0E-10	1.2E-10	6.8E-11	4.2E-11	3.4E-11
Effective Dose	8.9E-09	5.8E-09	2.9E-09	1.7E-09	1.0E-09	8.0E-10

GI-Tract Gastrointestinal Tract
St Stomach
SI Small Intestine
ULI Upper Large Intestine
LLI Lower Large Intestine

* In the biokinetic model for Np parameter values for the adult apply to ages > 25 y. For radioisotopes of this element the dose coefficients for the adult are based on the 50-y integrated doses following an acute intake at age 25 y.

References

ICRP (1989) *Age-dependent Doses to Members of the Public from Intake of Radionuclides: Part 1*, ICRP Publication 56. Pergamon Press, Oxford.

ICRP (1991) *1990 Recommendations of the International Commission on Radiological Protection*, ICRP Publication 60. Pergamon Press, Oxford.

FSC
www.fsc.org
MIX
Papier | Fördert gute Waldnutzung
FSC® C083411

Kolibri 360 GmbH
Ferdinand-Jühlke-Straße 7
99095 Erfurt, Deutschland
produktsicherheit@kolibri360.de